Solar Energy in Africa

Solar Energy in Africa
Possibilities & Capabilities

Editors

Mammo Muchie, Angathevar Baskaran
& Daniel Ayuk Mbi Egbe

AFRICA WORLD PRESS
TRENTON | LONDON | CAPE TOWN | NAIROBI | ADDIS ABABA | ASMARA | IBADAN | NEW DELHI

AFRICA WORLD PRESS
541 West Ingham Avenue | Suite B
Trenton, New Jersey 08638

Book design: Dawid Kahts
Cover design: Ananthy Baskaran

Cataloging-in-Publication Data may be obtained from the Library of Congress.

ISBNs: 9781569027325 (HB)
9781569027332 (PB)

Contents

Preface: Technological Capabilities in Solar Energy in Africa

Mammo Muchie & Angathevar Baskaran

Africa's population is rapidly growing at more than twice the global rate. In 2000, population of Africa was around 820 million and it accounted for about 13 per cent of total world population (6.1 billion). In 2018, this increased to around 17 per cent. Besides, Africa has the fast-growing young population. This is driving rapid urbanization, which over the last two decades has increased the urban population by 90%. It is expected that by 2040, an additional 580 million Africans will be living in urban areas, that is almost the current population of the European Union. As a result, demand for sustainable energy supply is also increasing rapidly. Growing Africa's energy demand is not only driven by rapid urbanization and expansion of economic productive activities, but also the need to provide access to energy supplies for a sizeable population which is currently living without such access. Over half of the population in Africa do not have access to electricity. Therefore, Africa is expected to emerge as a key driver of global energy demand growth in the coming decades. As the continent possesses abundant reserves of fossil fuels, hydropower, wind power, solar power and minerals, meeting such increasing energy demand is unlikely to pose serious supply problem. However, the challenge will be how to increase the share of renewable energy including solar energy, as the world is moving towards more clean energy. Furthermore, even though Africa has caused only 2 per cent of global energy-related carbon dioxide (CO_2) emissions, the continent is disproportionately affected by the effects of climate change. In this context, the role of solar energy and technology will be crucial. Solar energy is clean, safe, reliable and sustainable. Solar energy can help to reach people in remote areas and reduce dependence on fossil fuels and provide sustainable energy supply. The African solar energy market has enormous potential due to its abundant solar resource. It is also increasingly cost effective due to technological innovations and increasing economy of scale. Indeed, the governments across Africa are actively promoting the solar energy market.

It is important for Africa to embrace a vision of a clean energy future, with solar energy as a key contributor, as dependence on fossil fuels to meet the growing demand will cause serious damage to the environment, agriculture and public health. Global warming's harmful effects already can be felt in the African continent, and these effects have already surpassed the scientists' worst predictions. Therefore, using solar energy is important for sustainability in the long term, as it can replace use of fossil fuels to power homes, communities, businesses, and automobiles and will help mitigate global warming effects. Life-cycle analyses of solar photovoltaic (PV) systems show that they can radically reduce emissions of global warming pollutants compared with fossil fuels. For some PV technologies, life-cycle emission reductions are as high as 89 to 98 percent. Concentrating solar power plants also can drastically reduce fossil fuel use and emissions. For example, it was found that a concentrating solar power plant can generate the energy used to build the plant within the first five months in operation. There are many solar technologies that can play a role in Africa's energy system, and they can be used to help power the economy and community development.

Preface

Taking advantage of Africa's abundant solar energy potential would help transform socio-economic-environment development in Africa. But it will take sustained efforts by governments, businesses, universities, and communities to achieve this transformation. Governments need to employ creative new policy frameworks and clear roadmaps to efficiently reap the benefits of solar energy. It is imperative that governments lay the groundwork for building local technological capabilities in solar energy by adopting smart public policies.

This edited volume presents contributions from scholars covering various aspects of solar energy such as solar technology development, different solar applications, and government policies promoting solar energy.

Acknowledgements

We would like to thank the DST/NRF SARChI (Innoivation and Development) at the Tshwane University of Technology, Pretoria, South Africa for funding this research project

We recognize and appreciate the support given by Dr Jideofor Adibe, Adonis & Abbey Publishers Ltd, London, United Kingdom, which enabled publication of this book.

We also would like to thank the African Journal of Science, Technology, Innovation and Development (AJSTID) and extend our fullest appreciation and thanks to all the contributing authors.

We thank Jan B. Mwesigwa (www.janmwesigwa.com) our layout artist and Ananthy Baskaran, the cover designer, for their hard work and dedication to our book project. Finally, we would like to thank Divyaasiny R. Rajaghantham for her meticulous work in preparing this manuscript.

Chapter 1

Overview of Global Trends in Solar Energy*

Angathevar Baskaran[1], Mammo Muchie[2], VGR Chandran[3] & Daniel Ayuk Mbi Egbe[4]

* *Acknowledgement:* We would like to acknowledge the financial support provided by The RU Geran - Fakulti Program, University of Malaya under the Research grant no: GPF002P-2018.

1 Associate Professor, Department of Development Studies, Faculty of Economics and Administration, University of Malaya, Malaysia; Senior Research Associate of SARChI (Innovation & Development), Tshwane University of Technology, Pretoria, South Africa.Email: baskaran@um.edu.my & anga_bas@yahoo.co.uk

2 DST/NRF SARChI Professor (Innovation & Development), Tshwane University of Technology, South Africa; Technology and Management Centre for Development (TMCD),Oxford University, UK; Associate Faculty Professor, Sussex University, UK; Adama Science and Technology University (ASTU), Ethiopia. Email: MuchieM@tut.ac.za

3 Associate Professor, Department of Development Studies, Faculty of Economics and Administration, University of Malaya, Malaysia. Email: vgrchandran@um.edu.my

4 Professor, Institute of Polymeric Materials and Testing (IPMT), Johannes Kepler University Linz, Altenbergstr. 69, 4040 Linz, Austria. Member of Organic Electronics Association (OE-A), and a board member of the World University Service (WUS) in Germany. International Coordinator, African Network for Solar Energy (ANSOLE) (www.ansole.org). Coordinator of BALEWARE (Bridging Africa, Latin America and Europe on Water and Renewable Energies Applications) (www.baleware.org). Email: Daniel.egbe@ansole.org & daniel_ayuk_mbi.egbe@jku.at

It is no exaggeration to say that the Sun is the primary source of energy on Earth. It transforms hydrogen into helium through nuclear fusion, whereby energy is released as electromagnetic rays (photons). It is said that for every hour, the Sun provides the Earth with enough energy to satisfy global energy needs for a year. Solar energy currently accounts for about 0.1% of the global energy demand. However, solar energy can be the primary renewable source for meeting the global demand in the long term. For example, *Desert Power 2050 (DP2050)* sets out an ambitious plan to meet the energy challenges of Europe as well as the Middle East and North Africa (MENA), through an integrated renewable energy region called EUMENA. It states (Desertec Industrial Initiative (Dii), 2012, p.6):

> An integrated EUMENA power system allows Europe to meet its CO2 reduction targets of 95% in the power sector more effectively and more economically by importing up to 20% of its electricity demand from MENA. Europe thereby saves a total of €33bn. annually, or €30 per MWh of power imported from MENA. Meanwhile, desert power enables MENA countries to supply their own energy needs reliably from the abundant solar and wind resources in the region. MENA can thereby contribute to a 50% CO2 reduction in its power sector despite a massive increase in demand. At the same time, MENA benefits from an export industry worth up to €63bn. per year. Furthermore, Europe as well as MENA profit from a 40% drop in the marginal cost of CO2 emission reductions in the power sector.

The original Desertec initiative hoped to deliver 15% of Europe's power needs from solar plants in North Africa and the Middle East via undersea cables[1]. It envisaged to produce 125 gigawatts of electric power by 2050, that would be exported to Europe (after meeting the local demand) via high-voltage direct current cables under the Mediterranean Sea (McGrath, 2012).

According to the International Energy Agency (IEA), the global renewable energy sector is expected to grow rapidly in the coming decades. It is estimated that by 2026, 15-20% of the total energy supply globally are expected to come from renewable sources such as hydropower, geothermal energy, solar energy (*Solarthermics* and *Photovoltaics*), ocean tidal energy, wind energy and bioenergy (biomasses).[2]

According to the International Energy Agency (IEA), considering today's policy intentions targets and policy initiatives that have already been announced (i.e. 'The Stated Policies Scenario'), global energy demand will rise by 1% per year to 2040. More than half of this growth will be met by low-carbon sources, led by solar photovoltaics (PV), and solar PV will be the largest component of global installed capacity. Considering the "Sustainable Development Scenario", that takes into account of the measures needed to meet sustainable energy goals in full and are aligned with the Paris Agreement on climate change, almost all the growth in demand will be met by wind and solar PV. Increasingly, solar PV is witnessing large-scale deployment mostly in the power sector and its growth is showing that rapid change is possible, if the direction and speed are set by governments, as they can set the conditions that determine energy innovation and investment (International Energy Agency, 2019).

Solar PV has experienced a remarkable growth in 2017, as more capacities were installed globally than any other power generation technology. A total of 99.1GW of solar capacity was installed in 2017. To put this in perspective, solar capacity installed in 2017 almost equalled the total world solar capacity of 100.9 GW in 2012 (Solar Power Europe, 2018, p.5). Table 1 shows the prospects of global solar PV markets by 2022. China, India and the USA are expected to be the leading solar power markets in the world. China is expected to add 209 GW in the "Medium Scenario", followed by India with 78.4 GW, and the US with 62.6 GW (International Energy Agency, 2019).

In Africa, many nations, especially those around the deserts, receive a great deal of sunlight every day, creating an opportunity for the development of solar technology across the region.

This chapter provides an overview of global and African trends in solar energy in terms of technology, markets and industry.

Global Trends in Solar Energy

Solar Technology

Solar energy is derived from the photons of the sunlight, which can be converted into solar thermal energy or electrical energy. To harness photons of the sunlight, there are two types of solar technologies used: (i) PV (Photovoltaic) systems which use solar panels to convert sunlight directly into electricity; (ii) CSP (Concentrated Solar Power) that indirectly harvests the solar thermal energy to produce electricity (see Table 1). Photovoltaic technology converts sunlight directly into electricity. This is one of the fastest growing segments of the renewable energy industry and is well established in many countries around the world. Concentrated solar thermal power technology involves converting heat energy into mechanical energy through the use of a steam turbine, which is then used to produce electricity. CSP has some advantages compared to photovoltaic as it can readily incorporate thermal energy storage and hybridization to provide dispatchable power. It involves relatively 'low tech' manufacturing methods for solar collector fields that can be combined with available steam turbine technologies to achieve rapid scale-up. Globally, solar PV systems are used to generate electricity in most countries and continues to be the fastest growing power-generation technology in the world. Between 2004 and 2009, grid-connected PV capacity increased at an annual average rate of 60 percent (REN21, 2010).

Table 1: Solar Energy Technologies

Solar Photovoltaic (PV)	*Concentrated Solar Power (CSP)*
• It generates electrical power by converting solar radiation into direct current electricity using semiconductor. • Materials used for photovoltaic are monocrystalline silicon, polycrystalline silicon, and copper indium gallium selenide. • Photovoltaic cells are further segmented into inorganic and organic cells.	• It uses mirrors or lenses to concentrate a large area of sunlight or solar thermal energy onto a small area. • Electrical power is produced when concentrated light is converted to heat which drives heat engine connected to an electrical power generator. • CSP systems is sub-segmented into parabolic troughs, solar power towers, and others.

Source: REN21 (2009); Dutzik et al. (2010).

Large scale PV plants consist of one or more PV panels, a DC/AC power converter/inverter, racks, mounting fixtures, and electrical interconnections. They are used for electricity generation which is either stored, used directly for self-consumption, or is fed into large electricity grids. There are two grid-connected projects: (i) Ground-mounted PV; and (ii) Rooftop PV.

CSP has an advantage because its efficiency increases with irradiation level. This is not the case for solar PV, as its efficiency declines with rising collector temperatures. However, CSP systems involve higher investment costs compared to solar PV systems. Because of this, most solar installations are based on PV systems. They are placed in ground-mounted solar farms and on rooftops. However, most of the general solar electricity-generation-capacity is installed by CSP and large-scale PV. Both CSP and PV systems have been growing significantly since 2010.

A photovoltaic power station is a concentrated zone of development for solar power generation, which is also known as a solar park or solar farm. This is generally a large-scale PV system designed for supplying commercial scale power into the electricity grid.

Over the years, investment in renewable energy sources for electricity, heating and in biofuels has increased considerably, which led to more efficient technology both in terms of efficiency and cost. By 2010, nearly 11 GW of solar PV was produced globally (REN21, 2009; REN21, 2010). According to REN21 (2010), First Solar (USA) became the first firm ever to produce over 1 GW in a single year. Major crystalline module price declined by about 50–60%, from $3.50 per watt in 2008 to $2 per watt in 2009 (Chinese Renewable Energy Status Report, 2009).

Figure 1 shows the cumulative global solar PV deployments in different countries in 2018.

Figure 1: Cumulative global solar PV deployments by country in 2018 (509 GW-DC)

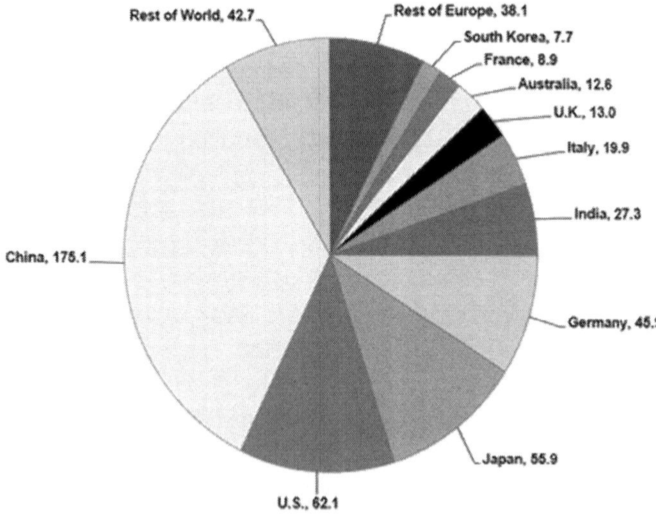

Source: Rhodes (2020)

By the start of 2018, global CSP installations reached 5.6 GW. Of this, a total of 400MW was added in 2017 from Morocco, China, and Kuwait. The majority of CSP was located in Spain and the US, but the market share of other countries has grown from 4% in 2012 to 22% in 2017 (NREL, 2018a, p. 21). Spain has submitted a plan to the European Union to install 5 GW of CSP between 2021 and 2030. These CSP plants would be designed to harvest and store during the day and produce power at night (complementing PV systems generating power). Heliogen, the CSP technology developer in the US, achieved temperatures exceeding 1,000º C by improving the reflectiveness of its tower arrays. This technology improvement could enable CSP plants to replace fossil fuel plants for industrial process heat (NREL, 2019, pp.18-19).

Solar Markets

Figure 2 illustrates the global solar energy market. Many countries, both developed and developing, are increasingly moving towards solar power, as it is more desirable, reliable, efficient, and environmentally friendly. The International Energy Agency estimated that 15-20% of the total energy supply contribution will be generated from renewable energy by the year 2026. By 2050, solar power is expected to become the main source of electricity in the whole world, if the capacity of solar PV systems can reach 4600 GW. Both India and China are expected to contribute about 50% of this capacity. In 2010, the capacity of global solar power was 50 GW. By 2019, it increased significantly to about 310 GW, of which 95 GW was added in 2017 alone, a 34% growth year-on-year." (Zafar, 2019).

It is argued that "the adoption of solar power will help improve the development of other sectors of the economy, such as the electronics industry, hence creating a lot of employment opportunities" (Salman, 2019). The effort towards the development of solar energy technology has been growing across all continents. Figure 2 illustrates the global solar energy market and Table 1 provides prospects of global solar PV markets.

In Europe, the UK, Germany and France are the leaders in solar power. In 2016, they have seen the growth of new solar installations measured at 29%, 21% and 8.3% respectively.

However, the growth of total solar power produced in Europe has been low. For example, the total solar power produced in Europe was about 100 GW in 2016, which increased to only 105 GW by 2019. The solar industry in Europe believes that the European Union needs to set a 35% target for renewable energy growth in order to sustain the industry (Zafar, 2019).

Germany has been one of the leading countries in solar power in the world. Since Germany introduced the Renewable Energy Act in 2000, a number of German companies became leaders in solar technology. However, their growth slowed down after 2012, forcing many companies to go out of business. Yet, Germany remains very strong in solar R&D and innovation. In 2017, Germany installed 1.76 GW of solar capacity, and about 25% of this was ground-mounted systems (750 kW). The rest were evenly distributed between the residential, commercial and industrial installation (Solar Power Europe, 2019, pp. 59-60).

Figure 2: Global Solar Energy Market

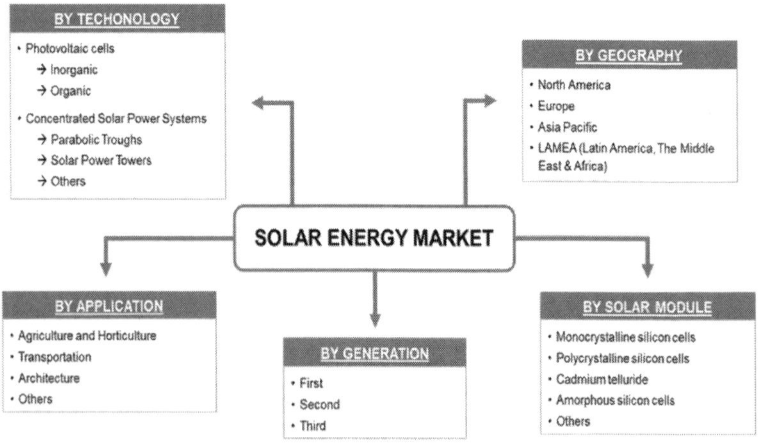

Source: Allied Market Research (2014)

Even though the growth of solar power is slowing in the US, it was still the second-largest PV market in 2017. The US solar market installed 10.6 GW of solar PV capacity in 2017, which was 30% less than that of 2016. However, it was still 40% more than the 2015 levels. Of the total installed capacity, 59% came from the utility PV segment, while distributed solar accounted for 41%. In 2017, all segments (residential and utility PV) experienced slow growth for the first time since 2010. However, the non-residential PV segment, which has been facing issues for a long time, has shown growth in 2017 (Solar Power Europe, 2019, p. 50). In the first nine months of 2019, the United States installed 7.1 GWDC of PV, a 10% increase over the same period in 2018. In 2019, California, Florida, the rest of the Southeast, and the Southwest, each installed more than 1 GWDC and the total solar capacity in the US was 55.4 GWAC, of which 32.9 GW were utility-scale PV and 22.5 GW were distributed PV (NREL, 2019, pp. 25-26)

By 2019, the US had deployed over 2 million solar PV systems (71,300 MW of solar capacity), generating over 100 TWh of electricity. In 2018, about 1.5% of total electricity was generated by solar. California is the leading state with over 26,000 MW solar PV capacity, followed by North Carolina. The total solar PV capacity is expected to more than double over the next five years, and California, Texas, Florida, Nevada, and North Carolina are projected to install about half of all solar PV capacities in the US. It is also expected that solar will be the fastest growing renewable energy sector in the US in the next decade (Rhodes, 2020). Table 1 shows the prospects of global solar PV markets.

Table 1: Prospects of Global Solar PV Markets

Country	2017 Total Capacity (MW)	2022 Total Capacity Medium Scenario by 2022 (MW)	2018 - 2022 New Capacity (MW)	2018 - 2022 Compound Annual Growth Rate (%)
China	130,751	339,751	209,000	21%
United States	51,527	114,170	62,643	17%
India	19,047	97,431	78,384	39%
Japan	49,251	78,551	29,300	10%
Germany	42,973	63,237	20,264	8%
Italy	19,392	26,924	7,533	7%
Australia	7,308	22,969	15,662	26%
France	7,999	19,702	11,703	20%
Mexico	1,174	15,252	14,078	67%
United Kingdom	12,676	14,742	2,065	3%
South Korea	5,480	14,497	9,017	21%
Spain	5,627	14,460	8,833	21%
Turkey	3,420	14,320	10,900	33%
Egypt	80	11,750	11,670	171%
Netherlands	2,681	11,430	8,750	34%
Brazil	1,147	8,442	7,295	49%
Taiwan	1,733	8,386	6,653	37%
Saudi Arabia	19	7,070	7,051	227%
Thailand	2,697	6,159	3,462	18%
Pakistan	894	5,787	4,892	45%

Source: Solar Power Europe (2019), *Global Market Outlook for Solar Power / 2018 – 2022*, P.21.

In 2016, over 66% of the global solar power was installed in Asia, of which half was installed in China. In 2017, China installed more renewables in a year than traditional thermal plants and the solar sector experienced exceptional growth. It installed 53.06 GW of solar PV (54% year-on year growth), increasing its total capacity to 130 GW. In 2017, solar contributed 42% to new generation capacity in China (53 GW) and 7% of cumulative capacity (130 GW). In 2017, for the first time, wind and solar contributed over half of all new electric generation in China (55%) (NREL, 2018, p.13). Chinese PV market growth represented approximately all global market growth in 2017 and over 50% of market growth in 2015 and 2016 (NREL, 2018a, p. 18). The "13th Five-Year Plan for Solar Energy Development" issued by the National Energy Administration (NEA) in 2016 set the target of 105 GW for installed capacity of solar PV by 2020. However, by the end of 2017, this target was exceeded as the installed capacity reached 130.5 GW, an increase of 69% over the same period of the previous year. In the first quarter of 2018, China installed 9.65 GW, a 22% year-on-year growth (Solar Power Europe, 2019, pp. 48-49).

India has been setting up solar parks across the country to meet its Paris Agreement commitments. In 2015, India set the target of achieving 175 GW of renewable energy capacity by 2022, including 100 GW of solar capacity. Rooftop PV is targeted to contribute 40% to the total solar. Since 2014, the Solar Power Corporation of India has approved 34 solar parks in 21 states to install solar power generation capacity of 20,000 MW (Solar Energy Corporation of India, 2020). In 2017, India had set a record for solar PV, as cumulative installed capacity exceeded 19 GW, with net yearly additions of 9.63 GW (127% market growth from 4.25 GW in 2016).

In 2017, solar was the largest source of new power generation in India with a 45% share (Solar Power Europe, 2019, p. 52). By 2019, India has installed 31 GW of solar capacity with an additional 17GW under construction (Singh, 2020). Solar power accounts for 9.6% of India's total installed power capacity. According to the Central Electricity Authority (CEA), India's solar PV parks boosted their output by 27% year-on-year in 2019, producing a combined 46,033 MWh. However, it was lower than the 69% growth seen in 2018 (Renewables Now, 2020).

According to the energy outlook report published by the Ministry of Economy, Trade and Industry (METI) in 2015, Japan's cumulative installed PV capacity in 2030 will be 64

GWAC (74 GWDC) and it will reach around 100 GWAC (120 GWDC) in 2030 and 200 GWAC (240 GWDC) in 2050. In 2015, Japan added 9.8 GWAC (equal to 10.8 GWDC). Since 2016, the Japanese PV market has been declining mainly due to grid constraints. The Japan Photovoltaic Energy Association (JPEA) estimates that around 6.0 GWAC (7.2 GWDC) was installed in 2017, resulting in a cumulative installed PV capacity of 49.2 GWDC (Solar Power Europe, 2019, p. 54).

Most ASEAN countries have introduced new policies to speed up the growth of solar. Vietnam has witnessed a rapid growth of the solar installation since 2017, and it has reached 4.5 GW by mid-2019, exceeding Australia's capacity, as well as the rest of ASEAN countries' combined capacity. In September 2019, ASEAN's largest solar farm began operation in Tay Ninh, in southern Vietnam, with the capacity of 420 MW. Thailand has launched the new Power Development Plan for 2018-2-37, which aims to install 10 GW of solar household rooftops by 2037 (Shani, 2020). Between 2014 and 2018, Asia has seen the installation of another 37 GW in PV capacity.

Latin America is expected to make a significant contribution to the continuous global solar capacity expansion. In 2018, solar power capacity in Mexico was 2,555 MW. Aura Solar I, in La Paz, Baja California and Villanueva Solar plant (310 MW) are among the largest solar power plants in Latin America. Brazil is the second largest producer of solar power in the region with 2296 MW. Chile is also a leading solar energy market in Latin America. By 2018, total solar capacity in the country increased to 2137 MW compared to 11 MW in 2013. The Atacama Desert solar plant, which is expected to generate 196 MW is the largest in South America (Chakrabarti, 2019).

According to the International Renewable Energy Agency (IRENA), the solar PV capacity in Latin America and the Caribbean could rise from 7 GW in 2019 to more than 280 GW by 2050 (IRENA, 2019). Table 2 shows the share of world GW-scale solar capacity newly installed in 2017 and 2018 respectively. The total newly installed global solar power capacity in 2017 was 9 GW and 14 GW in 2018. China, India, USA, and Japan are the leading GW-scale solar markets in the world. In

Africa, Egypt is emerging as the leading GW-scale solar market.

Table 2: Share of World GW-Scale Solar Markets, 2017 and 2018

Country	2017 (%)	2018 (%)
China	53	38
India	10	10
USA	11	10
Japan	7	7
Australia	1	3
Germany	2	3
Mexico	-	2
Brazil	1	2
Turkey	3	2
South Korea	1	1
France	-	1
Netherlands	-	1
Taiwan	-	1
Egypt	-	1
Rest of the World	11	18
Total Market Size	9 GV	14 GW

Source: Solar Power Europe (2019), *Global Market Outlook for Solar Power / 2018 – 2022*, p.47.

Table 3 shows details of selected major solar projects in different countries across the world. It is evident that China, India, US, and Dubai have installed or are currently in the process of installing several mega solar parks/farms. In Africa, Egypt and Morocco are the leading countries in large-scale solar projects.

Table 3: Major solar projects across the world

	Country	*Project*	*Capacity*	*Other Details*
1	Dubai	Mohammad bin Rashid Al Maktoum Solar Park	5,000 MW By 2030	1st Phase (2013): 13MW PV solar 2nd Phase (2017): 200MW PV solar 3rd Phase (2018-2020): 200 MW(2018), 600 MW (2020)
2	India	Kurnool Ultra Solar Park, Andhra Pradesh	1,000 MW	Largest operational solar power station in 2017
3	India	Kamuthi Solar Power Project, Tamil Nadu	648 MW	Largest solar power plant at a single location, Covers an area of $10KM^2$
4	India	Pavagada Solar Park, Karnataka	2000 MW	40 blocks each contributing 50 megawatts
5	India	NP Kunta Ultra Mega Solar Park, Andhra Pradesh	1500 MW	First phase was commissioned in 2016 (200 MW)
6	India	Rewa solar plant Madhya Pradesh	750 MW	Began supplying electricity in 2018
7	China	"Great Wall of Solar" in Tengger desert, Zhongwei, autonomous region of Ningxia	1,547 MW, started in 2012	Largest photovoltaic park in the world. 45 interconnected projects.
8	China	Datong Solar Power Top Runner Base	1GW Phase I completed.	3GW total capacity in 3 phases.
9	China	Longyangxia Dam Solar Park Qinghai Province	850 MW	Installed 300 hectares of solar panels above a fish farm
10	China	Yanchi Ningxia Solar Park	1000 MW	World's biggest continuous solar PV array when opened in 2016
11	China	Hangzhou Fengling Solar-Fish Farm in Cixi	200 MW	Completed in 2017

12	USA	Ivanpah Solar Power Facility, Mojave Desert. Base of Clark Mountain in California.	392 MW	Largest concentrated solar power facility in the world when it opened in 2014
13	USA	Solar Star projects Kern County California	586 MW	Completed in 2015. Comprise of 1.7 million photovoltaic modules
14	Egypt	Benban Solar Park	1465 to 1800 MW	Total 32 stations
15	Mexico	Enel Green Power-operated park	754 MW	2.5 million solar panels, First phase inaugurated in 2018
16	Morocco	Noor-Ouarzazate Solar Power Station	580 MW	Spread across 3,000 hectares
17	Vietnam	Tay Ninh	420 MW	Largest solar farm in ASEAN

Source: CNN (2020); Sun Powered Dreams (n.d.); (Shani, 2020).

Global Solar Industry

Based on existing technology, the global solar energy industry is bifurcated into photovoltaic cells and concentrated solar power systems (CSP). Photovoltaic cells are further segmented into inorganic and organic cells, whereas CSP systems are sub-segmented into parabolic troughs, solar power towers, and others. According to the solar module technology, the market is categorized into monocrystalline silicon cells, polycrystalline silicon cells, cadmium telluride silicon cells, amorphous silicon cells, and others.

In the global solar value chain, North America and Europe have largely focused on researches to maximize the solar potential. Asian countries such as China and Japan have significantly increased the production of solar technologies as a result of the governmental tariffs and merger and acquisition of local manufacturers. Solar technologies have also improved in the Middle East and Africa due to the increase in applications of solar energy for power generation, agriculture, and architecture.

Both developed and developing nations are increasingly looking at moving away from nuclear power and fossil fuels and towards solar, wind and other renewables. The global solar energy industry is expected to reach $422 billion by 2022 from $86 billion in 2015, growing at a CAGR of 24.2% from 2016 to 2022 (Allied Market Research, 2014).

Solar photovoltaic power is one of the fastest growing sectors in 2017, mainly driven by the falling prices and solar deployment boom in China. China is currently the world's largest solar energy market in terms of solar energy production and consumption. China's solar capacity of 53 GW in 2017 was the major contributor to the global solar energy capacity growth. Since 2012, China's share of global PV demand has grown from 10% to over 55%. Moreover, many Chinese solar energy companies have become global leaders.

According to market research, in 2015, the total global shipment of solar PV amounted to 50.8 GW, with China contributing 48%, Taiwan 20%, Malaysia 12% and Japan 6%. Malaysia aims to emerge as the world's second-largest PV producer by 2020, by increasing the current output of 12% of global production to 20% (Malaysian Reserve, 2017). Malaysia aims to benefit from the spillover effects of growing solar power usage worldwide and has been making efforts to develop the necessary solar ecosystem to obtain a market share of the growing solar power business both globally and within the region.

The Malaysian solar industry ecosystem comprises of 250 companies involved in both upstream and downstream activities. According to Malaysian Investment Development Authority (MIDA), in 2016, exports and local sourcing activities amounted to RM11.1 billion and RM1.42 billion respectively (Lee, 2017). According to Datuk Azman Mahmud, chief executive officer of MIDA, the National Renewable Energy Policy and Action Plan aims to achieve 2,080 MW of renewable energy by 2020. In 2017, Malaysia launched the Malaysian Solar

PV Roadmap 2030 to drive the country's solar PV industry forward, which is ranked as the world's third largest producer of PV cells and modules.

The global solar industry consists of many leading solar firms including: JA Solar (based in Shanghai, China), Canadian Solar (based in Ontario, Canada), Risen Energy (Ningbo, China), Zhongli Talesun (Suzhou, China), Jinko Solar (Shanghai, China), Trina Solar (in Changzhou, China), LONGi Solar (in Xian, China), Hanwha Q Cells (Seoul, South Korea), SunPower (Silicon Valley, USA), GCL-SI (Hong Kong), First Solar (Tempe, Arizona, US), Tongwei Solar (Chengdu, China), Motech Industries (Tainan, Taiwan), Aiko Solar (Guangdong, China), and Yingli Solar (based in Baoding, China). It is clear that several Chinese firms have emerged as global leaders in the solar industry. Table 4 provides the profiles of selected leading global solar firms.

Table 4: Profiles of Leading Global Solar Firms

Company	Headquarters	Operation/Markets/ Highlights
Jinko Solar	Shanghai, China	Shipped a total 11.4 GW of modules in 2018. The largest solar panel manufacturer in the world.
Canadian Solar	Ontario, Canada	Production facilities in China, Indonesia, Vietnam and Brazil. Shipped more than 32 GW PV modules over its 18-year history.
Risen Energy	Ningbo, China	R&D, production, sales and service of solar modules, products of photovoltaic technology, solar terminal application and integration. Exported to more than 30 countries.
Zhongli Talesun	Suzhou, China	Annual capacity of 4 GW cells and 5 GW modules. Cumulatively developed more than 6 GW PV power plants and its accumulative module shipment surpassed 10 GW.
JA Solar	Shanghai, China	Largest manufacturers of high-performance PV products. As of 2017, JA Solar's accumulative shipments reached 27 GW.
Trina Solar	Changzhou, China	Manufactures a huge range of solar modules for residential, commercial and utility scale installations. Largest supplier of solar equipment in India.

LONGi Solar	Xian, China	World's leading mono-crystalline solar module manufacturer and LCOE (levelized cost of electricity) solution provider.
Hanwha Q Cells	Seoul, South Korea	R&A based in Thalheim, Germany. Top solar panel suppliers in European and Japanese markets.
SunPower	Silicon Valley, US.	Solar panel manufacturers for over 30 years. Leader in solar innovations. Total solar energy generated has surpassed 18,000,000 MWh in 2019.
GCL-SI	Hong Kong	As of 2018, GCL-SI supplies nearly 30% of the total global wafer capacity wafer yield, represents 22.2% silicon market shares.
First Solar	Tempe, Arizona, US	Holds world records for thin film cell conversion efficiency (22.1%) and module conversion efficiency (18.6% aperture). Became the first company to lower its manufacturing cost to $1 per watt in 2009.
Tongwei Solar	Chengdu, China	New 30GW solar cell manufacturing hub in Chengdu would be built over a five-year period. Expects to reach a cell capacity of 60GW in 2022 and expand capacity to between 80GW and 100GW in 2023.
Motech Industries	Tainan, Taiwan	Became the world's largest professional manufacturer of photovoltaic cells with 3.2GW production capacity in 2016.
Aiko Solar	Guangdong, China	Production capacity of 4GW. In 2017, announced plans to expand cell capacity to 8GW in the near-term.
Yingli Solar	Baoding, China	Photovoltaic module capacity is 4 GWs. Distributed more than 15 GW solar panels to customers worldwide.

Source: Technavio (2019)

From Table 4, it is clear that Chinese firms play a leading role in the global solar manufactory industry and technology.

The National Renewable Energy Laboratory, Colorado (US), tracks quarterly PV shipments from some of the major solar companies. It reported that in Q3 2017, PV shipments fell by 20% from the previous quarter, but that was 31% more than Q3 2016 (National Renewable Energy Laboratory (NREL, 2018, page 45). Global PV shipments in 2017 were approximately 92 GW (i.e. 32% more than 2016). About 98% of these came from Asian countries, with China supplying 57% of the shipments. China and Taiwan are leading in terms of global c-Si wafer shipments. From 2004 to 2017, the compounded average growth rate of global PV shipments was 41%, which was mainly driven by Asian countries. In 2005, Japan, the US, and Europe shipped 92% of global PV modules, which fell to 5% in 2017. The US supplied approximately 0.5% of global PV modules in 2017 (NREL, 2018a, pp. 51-52)

In 2007, there was a significant geographic diversity of solar manufacturers. Since then, Asian manufacturers have become the dominant players. In 2017, for the first time since 2006, the First Solar (a US company) failed to appear in the top 10 companies for PV. However, First Solar more than doubled its production from Q3 2018 to Q3 2019, "as its Series 6 fleet efficiency, throughput, yield, and cost improved" (NREL, 2019, p.38). Despite increased tariffs in the US, imports of PV modules and cells have gone up significantly. In 2019, 18.6 GW of PV modules and 2.5 GW of cells were imported, that amounted to 21 GW available to the US market. Nearly 87% of PV module imports came from Malaysia, Vietnam, Thailand, and South Korea. In total, 96% of imports came from Asian countries and almost 50% of PV cell imports came from South Korea (NREL, 2019, pp. 39-40).

In 2017, all the top 10 PV manufacturers, individually, had shipped more than the entire industry did in 2007 (NREL, 2018a, p. 54). Table 5 shows the changes in global ranking of leading PV manufacturers (by shipments) between 2007 and 2017.

Table 5: Global Leading PV Manufacturers (by Shipments)

Rank	Manufacturer (2017)	Shipments (GW)	Manufacturer (2016)	Shipments (GW)	**Manufacturer (2007)**	**Shipments (GW)**
1	JA Solar	6.5	Trina	5.0	Sharp	0.4
2	Canadian Solar	5.4	JA Solar	4.9	Q-Cells	0.3
3	Zhongli Talesun	5.0	Hanwha	4.0	Suntech	0.3
4	Jinko Solar	4.9	Jinko Solar	3.9	Kyocera	0.2
5	Trina Solar	4.8	Motech	2.9	First Solar	0.2
6	LONGi	4.5	First Solar	2.7	Motech	0.2
7	Hanwha	4.2	Longi Lerri	2.7	Sanyo	0.2
8	Tongwei	3.8	Canadian Solar	2.4	SolarWorld	0.1
9	Motech	3.2	Yingli	2.4	Mitsubishi	0.1
10	Aiko	3.1	Shunfeng-Suntech	2.2	SunPower	0.1
Other		45.5		36.4		1.0
Total		91.9		69.5		3.0

National Renewable Energy Laboratory (NREL) (2018a), p.54.

Solar Energy Trends in Africa

The African energy demand is expected to grow twice as fast as the global average over the next two decades, due to fast growing urban population and industrial production. Africa's vast renewable resources can help to meet a major share of this demand. Particularly, Africa has vast solar resources that can be harnessed for both electricity generation and thermal applications. It can be utilised for small-scale purposes at household and community levels and large scales at industrial and national levels.

According to the International Energy Agency: "Africa has the richest solar resources on the planet, but has installed only 5 gigawatts of solar photovoltaics, accounting for less than 1% of global capacity. With the right policies, solar could become one of the continent's top energy sources" (International Energy Agency, 2019a, p.3). This is quite likely to happen as solar technology costs are declining rapidly due to the global solar market expansion which has created economies of scale.

Indeed, many countries have put in place policies and programmes to support the solar energy market. In South Africa, the government is actively promoting solar energy market through the Renewable Energy Independent Power Producer Programme (REIPPP). It is expected to be one of the leading countries in solar energy market in Africa. Kenya is also keen to produce more electricity from solar power parks by providing various incentives for using solar panels. It aims to achieve installed capacity of 200 MW of solar power by 2022 (PR Newswire, 2019). Also, many leading global solar firms are showing keen interest in developing the African solar market. For example, Sky Power has signed agreements to install 7 GW of solar PV capacity over five years in Egypt, Kenya and Nigeria.

Solar energy capacity has grown significantly in Africa in 2018. According to the International Renewable Energy Agency (IRENA), this growth was mainly driven by Egypt, South Africa, Kenya, Namibia and Ghana. They contributed 1,067 MW of newly installed PV capacity in 2018. Egypt has seen the fastest solar energy growth in Africa in 2018, by contributing around 581 MW of new PV capacity (fourfold increase compared to 2017 capacity), followed by South Africa which contributed 373 MW of PV capacity. South Africa is considered as one of Africa's largest solar energy markets, with an installed PV capacity of 2.5 GW. However, the growth has slowed down due to the postponement of its renewable energy auctions. Kenya has added 55 MW of PV capacity in 2018. However, solar power generated in the country is relatively smaller compared to hydropower capacity, which forms 65% of the country's energy needs. Namibia almost doubled solar PV capacity, from 46 MW in 2017 to 79 MW in 2018. Ghana added 25 MW of new solar PV installations in 2018 that increased the total cumulative installed solar energy capacity to 64 MW. Other African countries such as Nigeria, Botswana, Mali and Angola are also making strong efforts to increase their solar power capacity (ESI Africa, 2019). In March 2020, Benin secured a $21.1 million loan from the ECOWAS development bank to build solar arrays throughout the country. Burundi also obtained a $160 million donation from the World Bank to deploy PV systems (Publicover, 2020).

The "Solar Development in Sub-Saharan Africa Project" aims to promote the Regional Solar Parks in West Africa. Its objectives include: (i) regional solar integration, dispatch, and capacity building; (ii) variable renewable energy (VRE) production monitoring capabilities by creating a unit with required equipment, consulting services and training; (iii) system to share knowledge on planning, procuring and integrating solar generation into the grid; and (iv) regional solar generation projects such as Burkina Faso Regional Solar Park and Mali Regional Solar Park; and (v) raising grid investments (World Bank, n.d).

The solar energy market in Africa is far behind other regions of the world mainly due to the poor economic conditions and lack of infrastructure. Especially, the lack of funding and technical know-how are the major factors behind the slow growth of solar energy industry in many African countries (PR Newswire, 2019). However, it is widely recognized that solar energy can help facilitate a rapid and efficient growth of the energy sector in the African countries. Africa's cumulative solar PV installed capacity at the end of 2014 was 1,334 MW, which was more than ten times larger than in 2009 (127 MW). South Africa was leading this rapid growth, adding nearly 780 MW between 2013 and 2014. Kenya has also seen sizable investments in solar PV, with 60 MW installed by 2014. Deployment of CSP in Africa is small, as in the case of other regions in the world. Algeria, Egypt, Morocco and South Africa together have deployed six CSP projects by early 2015, with an installed capacity of over 180 MW. Projects for installing 6.4 GW have been underway in different countries including Botswana, Namibia, Sudan, and Tunisia. Among the 350 African solar PV projects in 2015, the majority have production capacities of between 10 MW and 100 MW. In some countries, much smaller systems (i.e. 1 MW to 10 MW) are also used at a utility scale, while large scale projects are also under construction (e.g. 500 MW in Namibia). As PV module costs fell by 75% from 2009 to 2015, the levelised costs of electricity (LCOE) for utility-scale projects in Africa have also fallen significantly. LCOE for African solar PV utility projects in 2013 and 2014 ranged between US$ 0.13 and US$ 0.26 per kilowatt-hour (/kWh). The cost for utility scale PV was the lowest in South Africa (US$ 0.075 per kWh), and the most competitive worldwide (IRENA, 2015, pp.14-15).

According to a study by LUT University in Lappeenranta, Finland, the West African countries should make solar PV the "prime source" in their future energy systems. This study backs a report by the International Renewable Energy Agency (IRENA) in 2018 by showing that West Africa has the potential to deploy up to 20 GW of solar by 2030. It expects that about 8 GW of PV capacity can be installed over a decade in the 15 ECOWAS countries, with Nigeria, Ghana and Côte d'Ivoire taking the lead (Publicover, 2020).

Contributions in the Book

This book presents the contributions by different scholars which discuss how solar energy is being developed and utilized in a variety of ways in different parts of Africa. They focus on the different aspects of solar energy in Africa, which are as follows:

- A Review of Solar Technologies for Buildings
- Design Simulation and Realization of an Intelligent Charge Controller (ICC) for Accumulators in Renewable Energy Power Stations
- Methodology of North-South Technology Transfer: Application to the Case of a Foodstuff Drier
- Reducing the Energy Bills of Higher Educational Institutions with Solar Energy
- Design of an Electronic Power and Charge Controller for Hybrid Energy Systems (EPOCHES)
- The Status of Solar Photovoltaic and Solar Thermal Applications in Malawi

- Situation of the Research on Renewable Energies in the Central Africa Region: The Case of Solar Energy
- Investigation of the Solar Energy Production and Contribution in South Africa
- The Current Status of Solar Energy Policy and Technology Development in Nigeria
- Kenya Solar Policy, Technology Development and Current Status
- ANSOLE DAYS: Demystifying Solar Energy Research through Artistic Creativity
- African Network for Solar Energy: Bridging Africans in Solar Energy Research

In Chapter 2 Noël Djongyang, René Tchinda and César Kapseu provide a review of solar technologies for buildings. They highlight that solar energy has been receiving attention in applying technologies and energy systems in recent years. Solar technologies for buildings relying on both passive and active systems are developed. Passive solar refers to those that absorb, store and distribute the sun's energy without relying on mechanical devices, while active systems are those where heat is transferred mechanically by the use of a working fluid such as air or a fluid that is typically water, or water-based. They also present some recent innovations in solar technologies for buildings. It is seen that today's solar architecture is undergoing a true revolution through the integration of renewable technologies into the fabric of buildings. Such systems are designed for heating, ventilation, thermal isolation, shading, electricity generation and lighting of buildings.

Patrice D. Dongo, Robert Tchitnga, Christian T. Tchapga, André R. Tchamda, Julius Kewir Tangka and Anaclet present in Chapter 3 their research on "Design simulation and realization of an intelligent charge controller (ICC) for accumulators in renewable energy power stations". Renewable energies contribute intensively to the protection of the environment in reducing the risk of global warming. However, their sources usually have irregular fluctuations, hence, they must be transformed and stored for permanent use. Batteries are indispensable entities for the storage of such energy produced by renewable energy power stations. A contribution is done through the innovative design of an intelligent device with multiple functions made using a microcontroller PIC16F877. Among others, the designed charge controller is capable of automatically detecting and regulating the temperature of the surrounding environment, which is a parameter that strongly influences the lifetime of batteries. It ensures the protection of batteries against the many disturbances that can occur during the charging and discharging processes. It also manages the use of the loads. This is what contributes to the rational use of the energy produced.

In Chapter 4, Yacoub I. Halawlaw and Mahamoud Y. Khaya present a methodology of north-south technology transfer for the case of a food stuff drier, as food stuffs are mostly used in dried form in Africa. This makes drying and driers a subject of major concern. Many works have been performed in this field. Is it necessary to make a new attempt to propose a solution? The authors argue that the approaches used till now are mostly based on the Western criteria of effectiveness. They explain why, in this case, the proposed technology is either simply rejected or misunderstood and misused. Furthermore, they show that it is possible to design technical objects which may fairly work in a Southern environment. They also propose a drawing of a solar drier, done using the CATIA software, and propose an approach that facilitates the design of a food drier.

César Kapseu, Noël Djongyang, Raoult T. Dabou, Maxwell T. Nsiéwé, Ibrahim Njukouyou and Fabrice Debel Eyono present the case of the University of Ngaoundere in Cameroon to show how energy bills of higher educational institutions can be reduced by using solar energy in Chapter 5. Humans rely heavily on energy in their daily activities and the demand is increasing. The conventional energy resources are limited and unequally distributed. Hence, a global requirement for sustainable energy provision becomes increasingly important. A feasibility study of electricity supply with solar photovoltaic energy at the University of Ngaoundere in Cameroon is presented. Evaluations showed that, for an estimated need of approximately 969 kWh/day, 3 162 solar panels (125 W, 24 V) are necessary to overcome the energy needs of the institution. The overall investment cost is estimated to be 153, 301, 485 US dollars (approximately 77 billion CFA). A simulation showed that the payback period of the investment is approximately nine years.

In Chapter 6, André R. Tchamda, Robert Tchitnga, Christian T. Tchapga and Martin Kom discuss the design of an Electronic Power and Charge Controller for Hybrid Energy Systems (EPOCHES). Existing devices for the management of electric storage modules in renewable energy power stations known as charge controllers are usually developed around the principle of controlling the voltage or current delivered by the generators for the charging of batteries. However, it is known that if the total power requested by different receivers (loads) connected to the batteries is greater than the power that can be provided by the batteries, there will be a rapid and deep damageable discharge of the batteries. The authors propose a module, which, apart from doing what existing charge controllers do, can also control the power delivered by the batteries to the loads.

In Chapter 7, Chifundo M. Tenthani, Esther Phiri and Thomas Senganimalunje explore the status of solar photovoltaic and solar thermal applications in Malawi, a Southern African country with approximately 85 % of the 13.6 million people living in the rural areas. At present, only 7.6% of the Malawian population have access to electricity. This figure accounts for 20% of the urban population and only 1% of the rural population. Solar energy presents a viable alternative because the country has good solar radiation resource of around 21 MJ/m^2/day for most times of the year. Although basic solar research is not well established in Malawi, there are efforts to rectify this by the government, institutions of higher learning, non-governmental organizations and to a lesser extent the private sector. Applied research mainly focuses on the adaption of PV systems developed elsewhere for local use with the aim of improving the quality of livelihoods. Traditional thermal applications of solar energy still dominate. Common applications in this category include passive space and water heating, clothes drying, raw and semi-processed food and fish drying. Opportunities exist for improvements in efficiency and hygiene by adopting technologies such as solar water heating and solar stills. New applications such as solar thermal cooking and power generation have good potential but require substantial investments. Current applications of solar PV include lighting, refrigeration, telecommunications and powering of low power electronic appliances. There are, however, several barriers faced in the penetration of solar; the main ones being high cost of capital equipment and high cost of spares.

César Kapseu, Noël Djongyang and Maturin Petsoko provide an overview of the current status of the research in solar energy in the central Africa region in Chapter 8. The achievement of sustainable development depends on reliable energy sources. In Africa, rural electrification poses some challenges. Autonomous energy resources such as solar, wind or hydropower are possible solutions. Africa has the greatest potential in solar energy worldwide. It appears that the research in the field of solar energy is promising. There are some research laboratories and private organizations working in the field, but the results are still under the expectations of the local populations. To boost these researches, it would be necessary to encourage and finance projects of regional integration. The Central African countries are aware of their potentiality in solar energy based on going projects.

In Chapter 9, Silas K. Mulaudzi, Mammo Muchie and Rudzani Makhado investigate the solar energy production and contribution in South Africa. Energy utilisation in South Africa is by far characterised by the high dependence on cheap coal which is also abundant. Coal utilisation for energy production makes South Africa to be the largest emitter of CO_2 in Africa. Renewable energy, such as solar energy technologies, has the potential to reduce the emission of trace gases to the atmosphere. South Africa has the best solar resources in the world; the country's average daily solar radiation varies between 4.5 and 7 KWh/m^2. The country has the potential of generating up to 24, 000 MW of CSP by 2030. This study seeks to establish the production, contribution and growth impediments of solar energy technologies in South Africa, mainly focusing on CSP and solar PV. We also critically review policy challenges with regard to solar energy production in South Africa. The findings show that the production of solar energy technologies is mainly hampered by weak institutional arrangement, strict regulatory framework, lack of skills, policy uncertainty and gaps as well as economic climate. Current contribution of solar energy technologies to energy security and climate mitigation effort is negligible due to the lack of CSP and PV plants. The distribution of solar energy across the country is spread in accordance with economic hub compared to solar energy potential. Areas around North West, Limpopo, and Mpumalanga have good solar potential but the application is relatively small.

Chapter 10 by Chigbogu G. Ozoegwu discusses the current status of solar energy policy and technology development in Nigeria. First, the author points out that it will be difficult to think of Nigeria meeting full electrification rate through grid expansion. Many terrains in Nigeria, like coastal areas and creeks, are too difficult or too isolated to be covered by grid power, leaving only the distributed power option. Moreover, it is not feasible to continue expanding the Nigerian grid because of insufficient national power generation capacity that frequently fluctuates from a relative high of around 3,700 megawatts to lower values and, sometimes, to total failure. Renewable energy, especially solar, is, therefore, the only option for a possible 100% electrification in Nigeria. Fortunately, Nigeria is geographically favoured for solar energy economy. Nigeria has a landmass of 923,768 km2 that receives daily solar irradiation of 3.5 kWh/m2 (145.83Wm-2) in the south and 7.5 kWh/m2 (312.50Wm-2) in the north. The solar energy resource is much above the global average ground-level solar irradiance of 170 Wm-2. The importance of solar energy for sustainable development of Nigeria has been identified and, as a result, many researchers have tried to assess solar energy availability in Nigeria. In this chapter, the author assesses the current status of solar energy policy and technology development in Nigeria.

Chapter 11 by Benard O. Muok assesses solar energy policy, technology development, and its current status in Kenya. The author highlights that Sub-Saharan Africa remains one of the regions with the highest energy poverty. Renewable energy has the potential to improve the generation capacity and contribute to the energy access. Driven by the global reduction in the cost of technology, solar is one of the fastest growing renewable energy technologies in the world, particularly in Africa. Kenya has one of the fastest and best-developed solar markets in sub-Saharan Africa. Total installation capacity was estimated at 20MWp with over 300,000 solar home systems installed in 2013. Despite these gains, the solar market in Kenya still faces major challenges to realize its full potentials. These challenges include slow or lack of local manufacturing capacity, inadequate financing mechanisms, business models which are still in a nascent stage and need more effort to make them sustainable, poor awareness especially in the rural areas, low technical capacity and inadequate policy framework including standards and quality control. The government of Kenya continues to provide policy environment to support the market growth, however, more still needs to be done in this aspect.

Chapter 12 presents "ANSOLE DAYS: Demystifying Solar Energy Research through Artistic Creativity" by Emelda Ngufor Samba. Finally, in Chapter 13, Daniel Ayuk Mbi Egbe briefly discusses "African Network for Solar Energy: Bridging Africans in Solar Energy Research".

Together, all the chapters underscore the critical importance of renewable energy in the African energy supply and the importance of developing specific and efficient technological cpabilities to harness solar energy. They also demonstrate the important role of solar energy in the socio-economic development of African countries and the need to build solar technological capabilities by combining knowledge, learning, innovation and competence building.

Endnotes

1 Desertec was set up in 2009 with a projected budget of 400bn euros to tap the enormous potential of solar and other renewables in North Africa to meet the demands of local region and Europe. However, due to political instability, cost and public subsidies involved, the project changed its strategies towards putting more emphasis on meeting the demands in MENA region first.

2 RENA21 (2009): *Renewables Global Status Report 2009 Update,* Paris: REN21 Secretariat.

References

Allied Market Research (2014), Solar Energy Market- Global Opportunity Analysis & Industry Forecast, 2014-2022. Available at: https://www.alliedmarketresearch.com/solar-energy-market

Chakrabarti, S. (2019), "Solar Power Statistics in Latin America: 2019", Solar Feeds, 19 August. Available at: https://solarfeeds.com/solar-power-statistics-in-latin-america/#Leading_Latin_Countries_in_Solar_Energy_in_2018-19

CNN (2019), "Global Gateway - Solar mega-projects around the world", 24 April. Available at: https://edition.cnn.com/2019/04/24/middleeast/gallery/global-solar-megaprojects/index.html

Dii (2012), *Desert Power 2050,* Munich, Germany: Dii GmbH.

Dutzik, T., Kerth, R., Sargent, R., and Del Chiaro, B. (2010), "Building a Solar Future Repowering America's Homes, Businesses and Industry with Solar Energy" Environment

America Research & Policy Center. Available at: https://environmentamerica.org/sites/environment/files/reports/Building-a-Solar-Future_0.pdf

ESI Africa (2019), "The Big 5 – Africa's fastest growing solar energy markets", Available at: https://www.esi-africa.com/industry-sectors/renewable-energy/the-big-5-africas-fastest-growing-solar-energy-markets-2/

International Energy Agency (2019), "Executive Summary", *World Energy Outlook*, Paris: IEA.

International Energy Agency (2019a), *Africa Energy Outlook 2019*, Paris: IEA. Available at: www.iea.org/africa2019

International Renewable Energy Agency (IRENA) (2015), "Africa 2030: Roadmap for a Renewable Energy Future". Available at: https://www.irena.org//media/Files/IRENA/Agency/Publication/2015/IRENA_Africa_2030_REmap_2015_low-res.pdf

International Renewable Energy Agency (IRENA) (2019), "Latin America and Caribbean on the Brink of Massive Solar Power Growth", 13 November 2019. Available at: https://www.irena.org/newsroom/pressreleases/2019/Nov/Latin-America-and-Caribbean-on-the-Brink-of-Massive-Solar-Power-Growth

Lee, J. (2017). A growing solar industry. The Star Online, 19 June. Available at: https://www.thestar.com.my/metro/smebiz/focus/2017/06/19/a-growing-solar-industry/

Malaysian Reserve (2017), Malaysia to be hub for solar cell manufacturing by 2030. 3rd April. Available at: https://themalaysianreserve.com/2017/04/03/malaysia-to-be-hub-for-solar-cell-manufacturing-by-2030/

McGrath, M. (2012), "Solar storm as desert plan to power Europe falters", *BBC World Service*, 19 November 2012. Available at: https://www.bbc.com/news/science-environment-20357167

NREL (National Renewable Energy Laboratory) (2018), "Q3/Q4 2017 Solar Industry Update", 18 February, NREL/PR-6A20-70917. Available at: https://www.nrel.gov/docs/fy18osti/70917.pdf

NREL (National Renewable Energy Laboratory) (2018a), "Q4 2017/Q1 2018 Solar Industry Update", May, NREL/PR-6A20-71493. Available at: https://www.nrel.gov/docs/fy18osti/71493.pdf

NREL (National Renewable Energy Laboratory) (2020), "Q3/Q4 2019 Solar Industry Update", 18 February, NREL/PR-6A20-76158. Available at: https://www.nrel.gov/docs/fy20osti/76158.pdf

PR Newswire (2019), "Analysis on Africa's Solar Energy Market, 2014-2019 & 2029", !3 December. Available at: https://www.prnewswire.com/news-releases/analysis-on-africas-solar-energy-market-2014-2019--2029-300974556.html

Publicover, B. (2020), "Solar emerging as best bet for West Africa's future energy mix", *PV Magazine*, 30 March. Available at: https://www.pv-magazine.com/2020/03/30/solar-emerging-as-best-bet-for-west-africas-future-energy-mix/

REN21 (2009), Renewables Global Status Report 2009 Update, Paris: REN21 Secretariat.

Renewables Now (2020), 11 February, Available at: https://renewablesnow.com/news/indias-solar-power-output-climbs-27-yy-in-2019-686657/

Rhodes, J. (2020), "The Future of US Solar Is Bright", Forbes, 03 February, Available at: https://www.forbes.com/sites/joshuarhodes/2020/02/03/the-us-solar-industry-in-2020/#20f1b45a5ed3

Salman Zafar, S. (2019), "Global Trends in Solar Energy Sector", *BioEnergy Consult - Powering a Greener* Future, September 21. Available at: https://www.bioenergyconsult.com/trends-in-solar-energy-sector/

Shani, N. (2020), "ASEAN Solar Sector: 2019 Highlights", *ASEAN Centre for Energy (ACE)*, 26 February 2020. Available at: https://aseanenergy.org/asean-solar-sector-2019-highlights/

Singh, M. (2020), "Solar power: In a slump India's booming", *Down To earth*. Available at: https://www.downtoearth.org.in/news/energy/solar-power-in-a-slump-india-s-booming-68445

Solar Energy Corporation of India (2020), "Statewise Solar Parks: Solar Park Implementation Status". Available at: https://seci.co.in/other-page/4/statewise_solar_parks

Solar Power Europe (2019), *Global Market Outlook for Solar Power / 2018 – 2022*, Amsterdam: Solar Power Europe. Available at: https://www.solarpowereurope.org/wp-content/uploads/2018/09/Global-Market-Outlook-2018-2022.pdf

Sun Powered Dreams (n.d), "The Largest Solar Plant in the World". Available at: http://www.sunpowereddreams.com/the-largest-solar-power-plant/

Technavio (2019), Top 10 Solar Energy Companies in the World 2019, 9 May. Available at: https://blog.technavio.com/blog/top-10-solar-energy-companies

World Bank (n.d), "Solar Development in Sub-Saharan Africa - Phase 1 (Sahel)", Available at: https://projects.worldbank.org/en/projects-operations/project-detail/P162580?lang=en

Chapter 2

A Review of Solar Technologies for Buildings

Noël Djongyang[1], René Tchinda[2] & César Kapseu[3]

1. Introduction

The recent commitment made by many countries to combat the harmful effects of fossil fuel energy has directed world attention on the implementation of policies geared towards an optimal energy performance and the use of renewable energies. The construction of buildings with sustainable energy systems necessarily plays an important role in such policies. Energy saving is also a high-priority worldwide. For this reason, energy-efficient measures are being increasingly implemented in all sectors (Kapseu *et al.*, 2012).

Buildings are a dominant feature in the modern society. We work, eat, sleep, and enjoy much of our leisure time indoors. Up to 80 % of an individual's life is spent in buildings (Djongyang, 2010). The residential sector accounts today for

1 Department of Physics, Faculty of Science, University de Ngaoundere, Cameroon. Email:noeldjongyang@gmail.com

2 University Institute of Technologies Fotso Victor, Bandjoun, University of Dschang, Cameroon, Email: ttchinda@yahoo.fr

3 Department of Process Engineering, National School of Agro-Industrial Sciences, University of Ngaoundere, Cameroon, Email: kapseu@yahoo.fr

about 40 % of the final energy consumption (Eicker, 2001; D&R International, 2009; Sadineni *et al.*, 2011). Most of this energy is used in heating, cooling, and in artificial ventilation systems (Schaeffer *et al.*, 2012). The energy used within buildings can be classified under two main categories (Kelly, 1998): high-grade (e.g. electricity, which can be efficiently converted to work via an electric motor) and low-grade (e.g. heat energy used to maintain conditions suitable for human comfort in buildings). Figure 1 shows the breakdown of energy usage in a typical air-conditioned office building (Kelly, 1998).

Energy use in buildings also accounts for nearly half of the carbon dioxide (CO_2) emissions in the world (BSRIA, 1999). More attention should therefore be paid to the type of energy systems used in buildings, which need to be sustainable and environmentally clean. Solar energy appears to be a good alternative to fossil energy. However, one of the major issues confronting users of solar energy-based systems is the relatively low efficiency of these systems compared to fossil fuel-based systems. Improving the efficiency of such systems has attracted authors for years, including passive, active, and hybrid solar technologies. In recent studies, Eicker and Dalibard (2011) developed a new photovoltaic-thermal system to produce electricity and cooling energy. Some works were also done in vernacular buildings, e.g. Shanthi Priya *et al.* (2012) conducted qualitative and quantitative analysis to investigate the indoor environmental condition of a vernacular residential building in coastal region of Nagapatinam, India. The results of this study showed that the solar passive techniques used in these vernacular residential buildings in coastal region provides comfortable thermal indoor environment irrespective of the outdoor climatic conditions.

Figure 1: The Energy Usage in a Typical Office Building

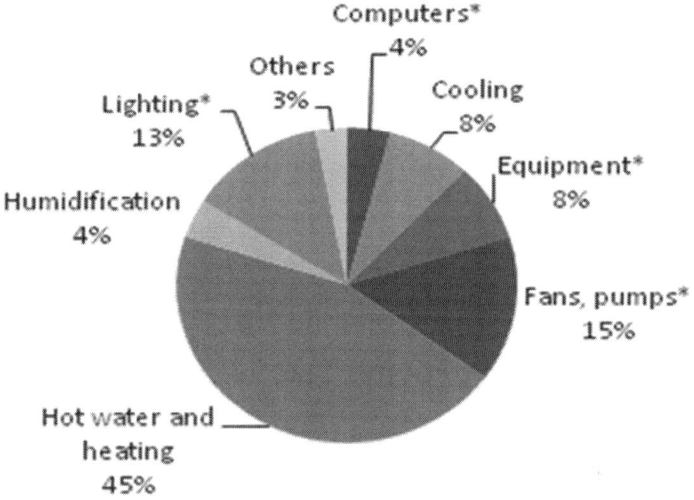

Many solar technologies have been developed in the past decades. The two main existing solar systems used in buildings are: passive solar systems, which refer to those that absorb, store and distribute the sun's energy without relying on mechanical devices; and active solar systems, where heat is transferred mechanically by the use of a working fluid such as air, or a fluid that is typically water, or water based. In recent technologies such as Building-integrated energy systems, the renewable technologies are integrated into the fabric of buildings. The recent innovations in the field of solar technologies for buildings will be explored below.

2. Solar Design Issues for Buildings

Building design can integrate several solar technologies including passive solar heating, ventilation air heating, solar domestic water heating and shading (Robertson and Athienitis, 2010). This section focuses on some of these technologies.

Advantages of Solar Energy in Building Design

Solar energy can substantially enhance building design. It offers several advantages compared to conventional energy: free after recovering upfront capital costs; payback time can be relatively short; available everywhere and inexhaustible; clean, reducing demand for fossil fuels and hydroelectricity, and their environmental drawbacks; can be building-integrated, which can reduce energy distribution needs.

Characteristics of Building Design Issues

Careful solar designs can (Robertson and Athienitis, 2010):

- Maximize possible solar transmission and absorption in winter to minimize or reduce to zero the heating energy consumption, while preventing overheating.
- Use received solar gains for instantaneous heating load and stores the remainder in embodied thermal mass or specially built storage devices.
- Reduce heat losses using insulation and windows with high solar heat gain factors.
- Employ shading control devices or strategically planted deciduous trees to exclude summer solar gains that create additional cooling load.
- Employ natural ventilation to transfer heat from hot zones to cool zones in winter and for natural cooling in the summer; use ground-source cooling and heating to transfer heat to and from the underground, which is more or less at a constant temperature, and utilize evaporative cooling.
- Integrate building envelope devices such as windows which include photovoltaic panels as shading devices or roofs with photovoltaic shingles; their dual role in producing electricity and excluding thermal gain increase their cost-effectiveness.

- Use solar radiation for day-lighting, which requires effective distribution into rooms or onto work planes, while avoiding glare.
- Integrate passive solar systems with active heating-cooling/air-conditioning systems in both design and operation.

Solar Technology Systems in Buildings

Solar buildings work on three principles: collection, storage and distribution of the sun's energy. These principles are integrated into two main solar systems: passive and active solar systems. They can simultaneously be integrated into building design.

Passive Solar Systems

Passive solar refers to systems that absorb, store and distribute the sun's energy without relying on mechanical devices like pumps and fans, which require additional energy. In other terms, passive solar involves no panels, no batteries, and nothing is fed into the general power grid. Passive systems could be used either for heating or cooling purposes. A passive solar building makes the greatest use possible of solar gains to reduce energy use for heating and, possibly, cooling by using natural energy flows through air and materials-radiation, conduction, absorption and natural convection (Robertson and Athienitis, 2010). It can optimize solar heat gain in direct heat gain systems, in which windows are the collectors and interior materials are the heat storage media. The principle can also be applied to water or air solar heaters that use natural convection to thermosiphon for heat storage without pumps or fans. According to Hibshman (1983), a good passive solar design should be as presented in Figure 2.

The three most common solar passive systems are (Hibshman, 1983):

- Direct gain system, which allows sunlight to windows into an occupied space where it is absorbed by the floor and walls.
- Indirect gain system, where a medium of heat storage such as wall, in one part of the building absorbs and stores heat, which is then transferred to the rest of the building by conduction, convection or radiation.
- Isolated gain system, where solar energy is absorbed in a separate area such as greenhouse or solarium, and distributed to the living space by ducts.

Figure 2: Diagram Showing Good Passive Solar Design.

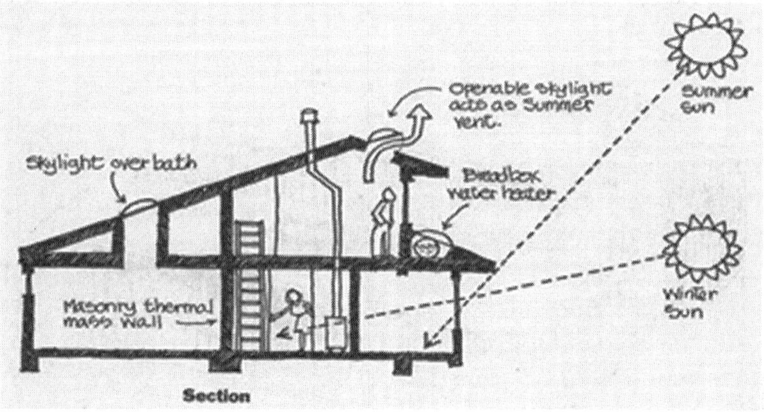

In general, for passive-solar architecture, the most significant design parameters which alter the solar contribution to the total cooling and heating load inside the building are aspect ratio of walls, orientation of the building, window details (size and location) and proper sun shade to control the amount of admission of incident solar radiation (Ralegaonkar and Gupta, 2010). To meet the heating and cooling requirements, the selection of proper building orientation for a particular aspect ratio should be done with the computation of beam radiation falling over exposed wall surface. Small scale modelling technique is one of the easy and best methods of experimentation to analyze the effectiveness of a particular system. For fast and accurate computation simulation, software is the best suitable option to predict and control the indoor environment. Sunlight area is a measure to determine the radiation interception, which regulates temperature inside the buildings. Graphical determination of sunlight area will be helpful for easy visualization and accurate measurement. Windows, which play vital role for solar incursion inside the buildings, should be shaded properly to regulate the sun entry as per seasonal requirements in composite climates. External static sunshades are most efficient among all the types of sunshades as they restrict the sun before it interacts with building components (wall, window). The existing external static sunshades (horizontal, vertical and egg create) satisfy shading needs partially. To check the effectiveness of a particular sunshade shading mask can be plotted. The efficient external static sunshade should be designed as per the location constraint, which will follow sun path at a particular geographic location whose shading characteristics can be controlled as per seasonal requirements.

The key to designing a passive solar building is to best take advantage of the local climate. The key aspects of passive solar design are interlinked and dependent to the following design parameters: location and orientation of a building; fenestration area, orientation and type; thermal massing and envelope characteristics; amount of insulation; shading devices-type, location and area; effective thermal storage insulated from the exterior environment, as well as amount and type; sensible such as concrete in the building envelope with exterior insulation, or latent such as phase-change materials. The major source of heat in a passive solar building is the light energy from the sun, which is converted into heat energy within the building. Apart from warming the inner spaces of the building during the day, this heat can also be stored for use when light is not available, notably at night. Maximizing the amount of light that enters a building in winter is fundamental to its ability to be warm then and reduce reliance on active forms of heating. Conversely, being able to limit the amount of light that enters in the Summer time is vital to preventing overheating at that time of year, thus reducing the need for air conditioning.

Building Orientation

The direction of a building and its major openings face is crucial to its functioning as a collector of light and heat (Kachadorian, 2006). The best way is to orient the house on the east-west axis and concentrate most of the house's glazing on the south wall. This allows the home to receive the most direct sunlight for the longest period of time (Hibshman, 1983; Kachadorian, 2006; Website A). Table 1 show the solar gain received according to the orientation of the building (Kachadorian, 2006).

Glazing

Glazing should be greatly reduced on the east and west walls and should be virtually eliminated on the north side of the home because most cold winds in winter come from the north and west (Desbarats, 1980; Website A). Table 2 presents coefficient of solar heat gain (ASHRAE Fundamentals, 1997).

Table 1: Solar Gain Received According to the Orientation of the Building

Orientation of building	Solar gain received
True south	100 %
22.5° away from south, either south-south-east or south-south-west	92 %
45° away from south, either south-east or south-west	70 %
67.5° away from south, either east-south-east or west-south-west	36 %

Table 2: Visible Light Transmission - Solar Heat Gain Coefficient (per cent)

Glazing system (6 mm glass)	Clear	Blue-green	Grey	Reflective
Single	89-81	75-62	43-56	20-29
Double	78-70	67-50	40-44	18-21
Double, hard low-e, argon	73-65	62-45	37-39	17-20
Double, soft low-e, argon	70-37	59-29	35-24	16-15
Triple, hard low-e, argon	64-56	55-38	32-36	15-17
Triple, soft low-e, argon	55-31	52-29	30-26	14-13

Thermal Mass

It is important to use building materials that have a large amount of thermal mass. Figure 3 shows how thermal mass works (Hibshman, 1983). The rudiments of solar passive design were developed and used through the centuries by many civilizations across the globe. Many of these early civilizations built dwellings that were better suited to their climatic surroundings than those built today in most developed and developing countries. Examples of houses built in traditional, indigenous styles are presented in Figure 4 (Website B).

Figure 3: How Thermal Mass Works

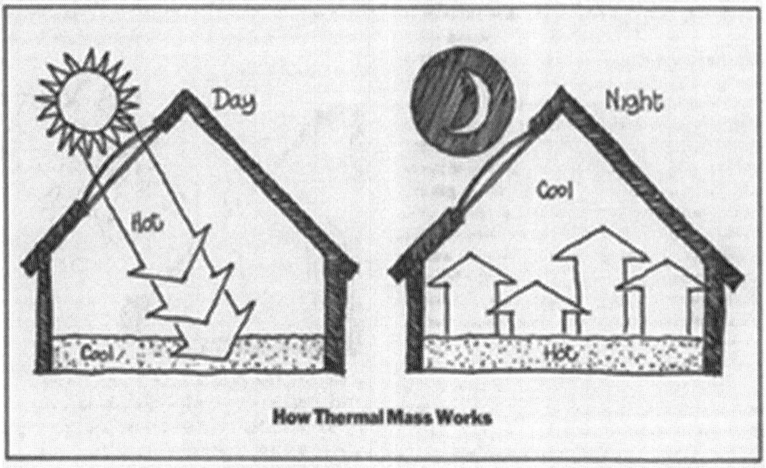

Figure 4: Traditional Home Structures Making the Most of Natural Resources

Passive Solar Heating and Cooling

Heating the building through the use of solar energy involves the absorption and storage of incoming solar radiation, which is then used to meet the heating requirements of the space.

Conversely, passive solar cooling systems function by either shielding buildings from direct heat gain or by transferring excess heat outside. Passive cooling is being employed as a low-energy consuming technique to remove undesirable interior heat from a building in the hot seasons. There are numerous ways to promote this cooling technique.

Active Solar Systems

In active system, heat is transferred mechanically by the use of a working fluid such as air, or a fluid that is typically water, or water based. An active solar system uses mechanical equipment to collect, store and distribute the sun's heat (Robertson and Athienitis, 2010). Active systems consist of solar collectors, a storage medium and a distribution system. Such systems are commonly used for water heating, space conditioning, producing electricity, process heat, and solar mechanical energy. In addition to solar collectors, other renewable energy systems like biomass to support the solar passive features as they allow a greater degree of control over the internal climate and make the whole system more precise.

Solar Collectors

The three main types are: flat plate collectors, commonly used for solar water heating systems in home and solar space heating; evacuated-tube collectors which can achieve higher temperatures; and integral collector-storage system also known as ICS or "batch" systems. The growth in the use of solar collectors for covering building loads (domestic hot water preparation and space heating mainly) has shown that these systems are mature and technically reliable. Research and development of solar technologies has led to very efficient solar collectors and systems (D'Antoni and Saro, 2012). The most diffuse solar technologies are flat-plate solar collectors and a wide range of typologies and materials are available on the market. The façades of buildings can be important solar collectors, and, therefore, become multifunctional. In addition, solar collectors can be used to enhance the appearance of the façade when considering their aesthetic compatibility. Currently, installation of collectors on the south tilted roofs, south walls, balconies or awnings of buildings are the feasible approaches for integration of solar collectors into buildings (Wang and Zhai, 2010).

Active Solar Heating and Cooling

Active solar heating main purpose is to use the solar cells to collect and absorb solar radiation. The radiation is converted into heat by solar cells and then transferred to the interior of the building. Flat-plate collectors are typically used for this purpose. Figure 5 shows example of active solar heating system (Website C).

Figure 5: Active Solar Heating.

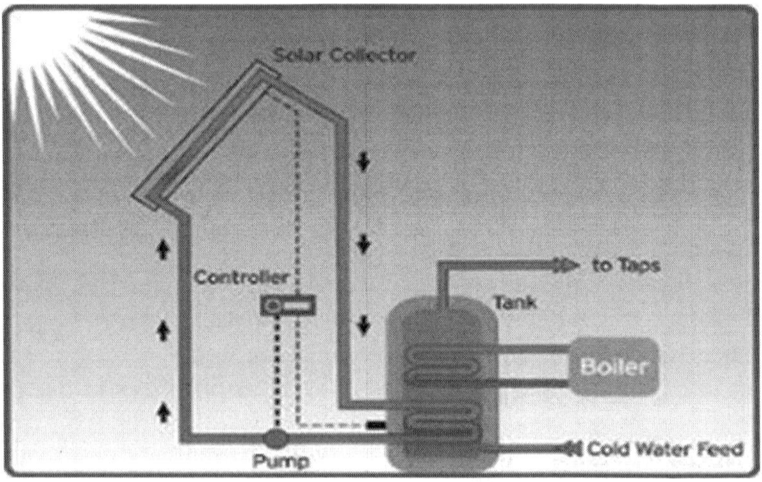

Active solar heating works in two ways:

- Solar liquid heating where water or antifreeze is warmed through the use of solar cells. Then the liquid is pumped into home and heats it via radiators.
- Solar air heating where solar panels warm the air then that air is pumped into home through a forced air pump.

Conversely, active solar cooling systems transfer a heated liquid from the solar collector to run a generator or a boiler activating the refrigeration loop which cools a storage reservoir from which cool air is drawn into the space.

Solar Electric System

Solar electric systems use photovoltaic cells to absorb solar radiation. These photovoltaic cells are semiconductors. Electrons are loose out when the cells absorb sunlight; this is called photoelectric effect. Then free electrons move to circuit where they form an electrical current. Cells are connected to form panels and each panel creates 10 to 300 watts. Usually a building or home has several panels for its power supply. The panels are fixed in a specific angle, or panels are having tracking device to maximize power for a day. Solar panels can be stand-alone or hooked up to the power grid. There are several solar array systems in the southwest that supply the electric grid.

Hybrid Solar Systems

Hybrid powers systems combine two or more energy systems or fuels that, when integrated, overcome limitations of the other, such as photovoltaic panels to supplement grid-supplied or diesel-generated electricity. Hybrid systems are the most common, except for the direct gain system, which is passive. Hybrid renewable energy systems are able to fill a number of roles:

- PV panels with integrated phase change material to maintain low cell temperatures thus improving efficiency while storing enough thermal energy for space or water heating.
- 'Hybrid renewable/energy efficient system' which enhance the effectiveness of the renewable technology being utilised. A renewable technology with a high coefficient of performance (e.g. ground source heat pumps) would

produce a synergistic effect when combined with a hybrid PV system.

- Hybrid energy systems can also benefit from being used in conjunction with passive applications. They are useful for applications in environments where daylight is restricted and there is perhaps poor ventilation also.
- Using natural ventilation with natural day-lighting, super-reflective chimneys or 'light pipes' with integral ventilation can be beneficent. Lighting levels can be doubled (Hill, 2002) and this effect can then be enhanced by coating the pipes with 'dichoric' coatings to absorb increased levels of solar thermal energy to improve performance of the stack effect. The stack effect draws in fresh air without the need for mechanical ventilation equipment.

3. Building-Integrated Energy Systems: Solar Façades Technologies

These are renewable technologies that are integrated into the fabric of buildings, displacing some of the traditional building materials, and drawing energy from its immediate environment (Dwyer, 2003). Building integrated renewable energy sources can either make use of its energy supply directly for electricity generation or heating, or passively in areas such as wind induced ventilation, transparent insulation or natural lighting. Integrating photovoltaic panels into a building façade represents a significant step forward in the application of this relatively new technology. Such a façade serves not only as a renewable source of electricity, but also as a source of heat for building heating and cooling.

Figure 6: Solar Façades Classification

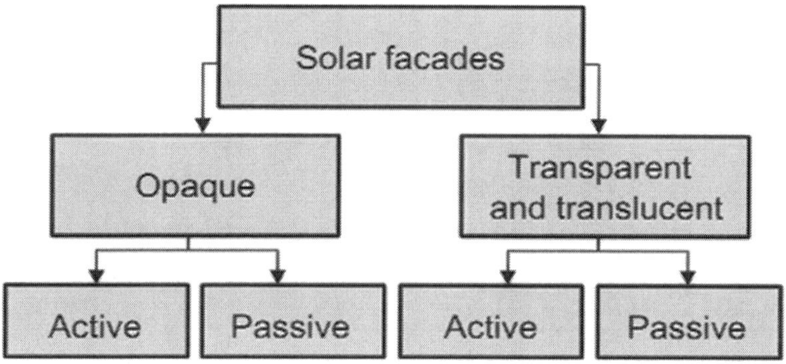

Today solar architecture is undergoing a true revolution because of the development among others, of solar façades designed for heating, ventilation, thermal isolation, shading, electricity generation and lighting of buildings. Figure 6 illustrates their general classification (Quesada *et al.*, 2012).

Opaque Active and Passive Solar Façades

The opaque solar façades can absorb and reflect the incident solar radiation but cannot transfer directly solar heat gain into the building (Quesada *et al.*, 2012). When such opaque solar façades transform the incident sunlight into electricity for immediate use or for transmitting the thermal energy into the building by the use of electrical or mechanical equipment (pumps, fans, valves, control equipment), then they are called opaque and active solar façades (Figure 7).

Figure 7: Opaque Solar Façades

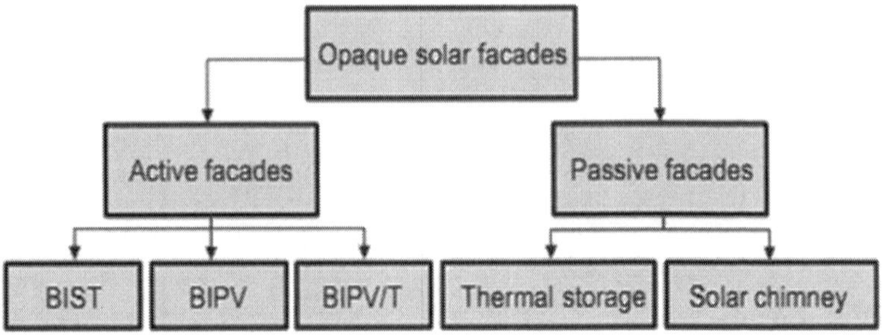

Active Solar Façades

Active and opaque solar façades include Building-integrated solar thermal (BIST) system, Building integrated photovoltaic (BIPV) system, and Building-integrated photovoltaic thermal (BIPV/T) system.

Building-Integrated Solar Thermal System

A building-integrated solar thermal system for façades can be conceived as the application of solar collection equipment to the façade of a building so that the equipment performs the function of an envelope while it simultaneously collects solar energy for heating purposes (Figure 8) (Quesada *et al.*, 2012).

Figure 8: Schematic Diagram of Building-Integrated Solar Thermal System

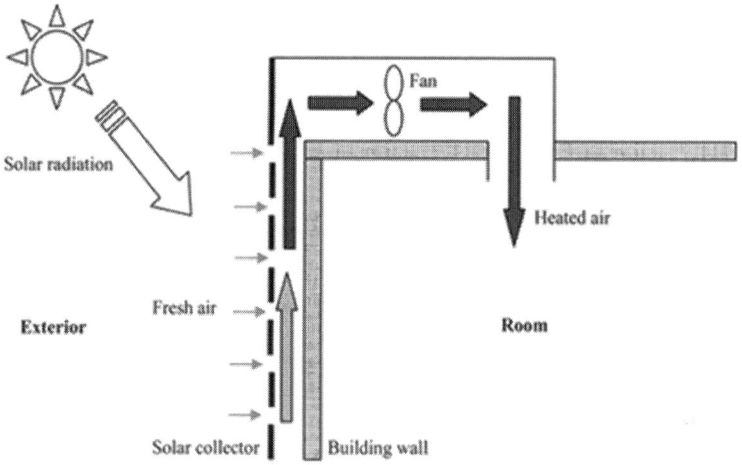

Building Integrated Photovoltaic (BIPV) System

Henemann (2008) describes a building integrated photovoltaic (BIPV) system as photovoltaic cells which can be integrated into the building envelope as part of the building structure, and therefore can replace conventional building materials, rather than being installed afterwards. BIPV (Figure 9) use photovoltaic surfaces that can be integrated with standard roofing, glazing or cladding products.

BIPV can be used in any external building surface. Rather than sticking out like a sore thumb, BIPV modules can be naturally blended into the design of the building, creating a harmonious architecture. Also, as an added benefit, air flow behind the solar cell reduces their temperature which improves their efficiency and longevity (Quesada *et al.*, 2012).

Figure 9: Schematic Diagram of Building Integrated Photovoltaic System.

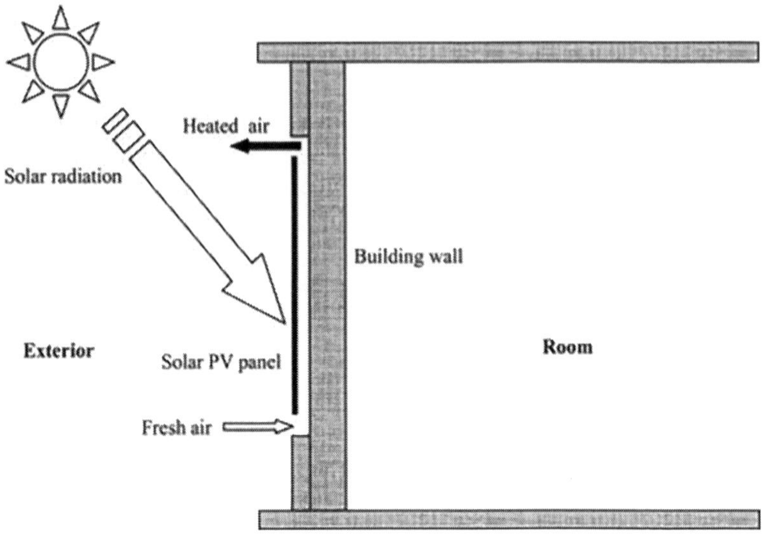

Building-Integrated Photovoltaic (BIPV) has been developed rapidly in the past decade. A number of BIPV projects are found worldwide. Currently, there is much development in PV roofing, PV shading elements and PV cladding or semi-transparent curtain wall components. With PV cladding it is best to have a vented cavity behind the panels so as to operate at lower temperatures. By following such a construction approach one may also develop an effective rainscreen system which hinders rain penetration. PV roofing is installed much the same way as conventional roofing and is available in shingles; tiles and metal standing-seam roofing. PV shading can be effective as a window shading element, entrance canopy or walkway shading. PV panels can be opaque, used where no light transmission is needed, or semi-transparent for areas where light is wanted, such as atriums or skylights, but some shading is needed to reduce cooling loads.

Sunshade PV claddings are typical applications of the BIPV systems to achieve an aesthetically pleasing outlook and energy efficiency in buildings (Fung and Yang, 2008). In addition to sunshade PV claddings, architects and building engineer s have tended to use semi-transparent BIPV modules to replace the traditional glazing in recent years for energy efficiency and aesthetic consideration. This kind of PV modules is usually composed of two layers of highly transparent glass sheets, and a serious of opaque solar cells that are encapsulated in between the two glass sheets (Figure 10).

By comparison with centralised PV plants, BIPV systems offer the "double dividend" of reduced economic costs and improved environmental performance. This double dividend is increased if the economic and energy costs of avoided cladding materials are taken into account (Oliver and Jackson, 2001).

Figure 10:A Typical Structure of a Semi-Transparent PV

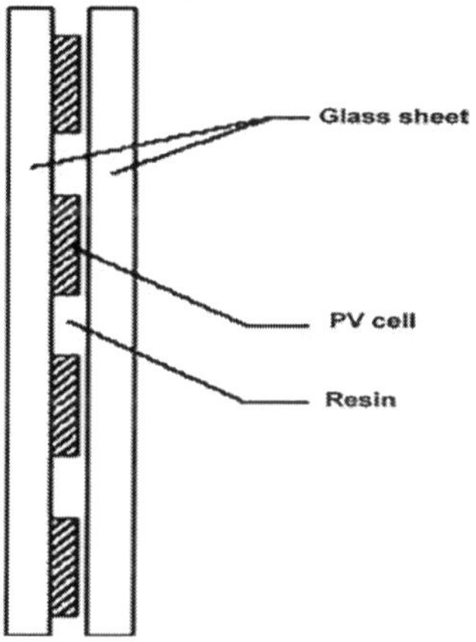

Building-integrated Photovoltaic Thermal (BIPV/T) System

A building integrated photovoltaic thermal (BIPV/T) system combines the functions of a building integrated photovoltaic system with those of a building-integrated solar thermal system. This combination seeks to achieve a most efficient use of a solar energy-collecting surface in terms of both an optimal electrical conversion and air/water heating (Figure 11) (Quesada *et al.*, 2012). Building integrated photovoltaic thermal (BIPVT) systems have been growing under rapid development over past decades and several buildings with BIPVT systems exist across the world.

Figure 11: Schematic Diagram of Building-Integrated Photovoltaic Thermal System (BIPV/T)

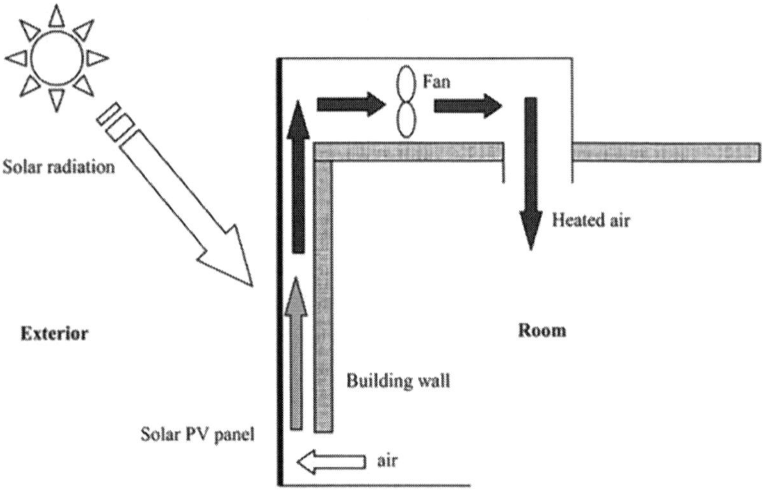

Building Integrated Concentrating Photovoltaics

For building integration, Concentrating Photovoltaic (CPV) systems can offer a host of advantages over conventional flat panel devices, the most notable being (Chemisana, 2011): a higher electrical conversion efficiency in the PV cells, better use of space, ease of re cycling of constituent materials, and reduced use of toxic products involved in the PV cells' production process. However, the viability of Building-Integrated Concentrating PV (BICPV) systems is dependent on their ability to offer a comparative economic advantage over flat panel photovoltaic technologies whose market prices are decreasing from day to day and which offer other advantage s such as ease of replacement of structural elements.

However, buildings Integrated Concentrating Photovoltaics (BICPV) need to be designed in such a way which minimizes costs allowing them to compete with standard flat panel technology, the manufacturing costs of which are falling continually. In addition to being technically and structurally sound, solar concentrators apt for architectural integration must fulfil the following requirements, these being a generalization of the criterion formulated by the IEA PVPS Task 7 workgroup for evaluation of the aesthetic quality of buildings integrated photovoltaics (Chemisana, 2011): Natural integration; Architecturally pleasing design; Good composition of colours and materials; Dimensions that fit the gridula, harmony and composition; Conformity to the context of the building; Well-engineered and innovative design.

BICPV systems may be installed either on the building façade or on the roof (which may be flat or sloped) producing in each case a different visual impact. Depending on the type of device, the system may be integrated in such a way that it is unseen, plays some role in the architectural aesthetic or that it constitutes in itself an architectural concept (Swanson, 2000).

Passive Solar Façades

Passive Solar Façades Include Thermal Storage Wall and Solar Chimney.

Thermal Storage Wall

Thermal storage wall combines the functions of solar collector and storage into a single unit. Heat is transferred from the wall to the room air and to the air between glazing and wall, by radiation and natural convection. Reducing indoor air temperature swings is one of its principal functions (Figure 12) (Quesada *et al.*, 2012).

Figure 12: Schematic Diagram of Thermal Storage Wall

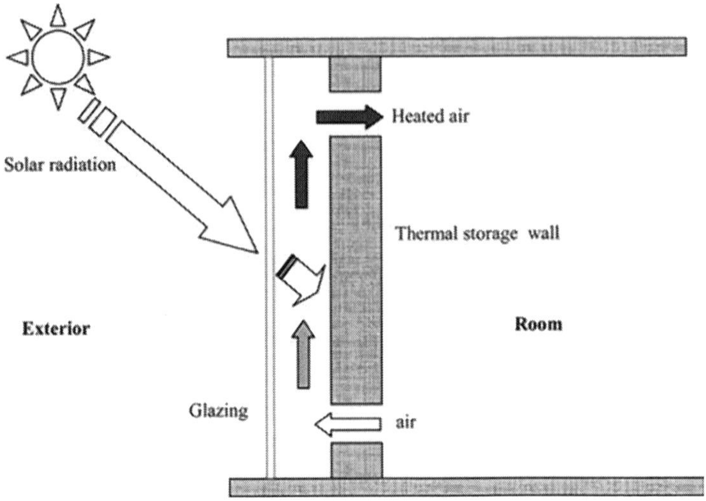

Solar Chimney

A solar chimney is a structure that consists mainly of one heat-absorbing glazed surface and it is constructed on the wall facing the direction of the sun. When solar energy heats the chimney and the air within it, it produces an updraft of air in the chimney. The natural aspiration created at the chimney's base can be used to ventilate the building (Figure 13) (Quesada *et al.*, 2012).

Figure 13: Schematic Diagram of Solar Chimney

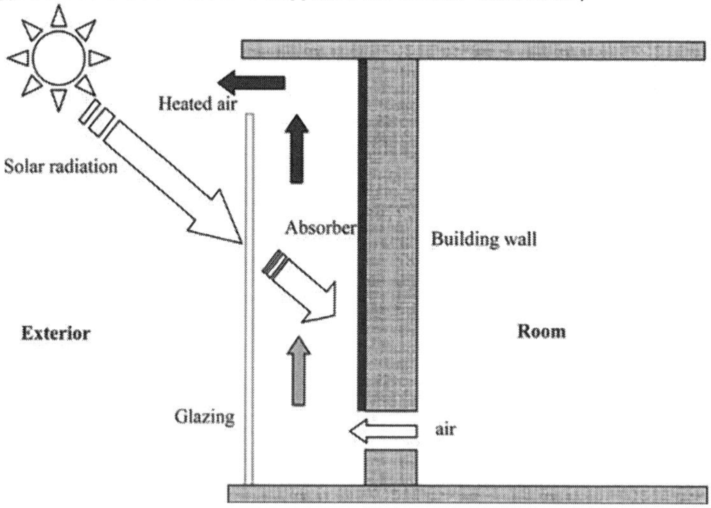

As a simple and practical bioclimatic design methodology, solar chimneys are receiving considerable attention for reducing heat gain and inducing natural cooling or heating in both commercial and residential buildings because of their potential benefits in terms of operational cost, energy requirement and carbon dioxide emission. In practical civil buildings, solar chimneys can be installed on the walls and roofs. For the purpose of improving natural ventilation performance and achieving better indoor thermal comfort, solar chimneys are always applied in the form of integrated configurations. Solar chimneys can also be used to combine with natural cooling systems so as to enhance the cooling effect inside buildings. Besides, active solar systems may be utilized to enhance the ventilation performance of solar chimneys (Zhai *et al.*, 2011). Figure 14 shows solar chimney configurations (Harris and Helwig, 2007).

Figure 14: Solar Chimney Configurations

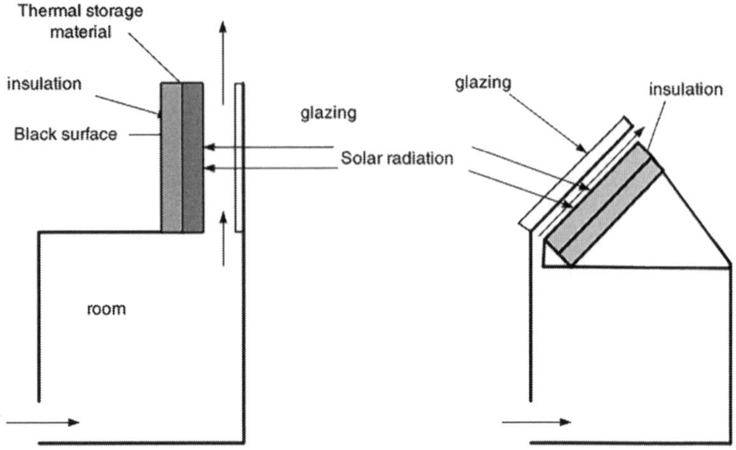

4. Example of Low Energy Building: Solar XXI Building

Solar XXI building is a low energy office building with 1500 m^2 of gross floor area located in Lisbon, Portugal (38° 46′ N, 9° 11′ W) (Marta *et al.*, 2011). It is a low energy office building where passive and active solar strategies have been applied to reduce the use of energy for heating, cooling and lighting, combining also an extensive photovoltaic façade for electricity production. Solar XXI opened in 2006, is considered as a highly efficient building, close to a net zero energy building (NZEB), where the difference between the energy consumed and that produced is 1/10th of the energy consumed by a Portuguese standard new office building. Its design includes many energy efficiency concepts, such as a high insulated envelope, south sun exposure, windows external shading, photovoltaic panels heat recovery, ground-cooling system, daylighting, stack effect and cross ventilation.

5. Conclusion

Some solar technologies for buildings are presented: passive solar systems which refer to those that absorb, store and distribute the sun's energy without relying on mechanical devices, and active systems where heat is transferred mechanically by the use of a working fluid such as air or a fluid that is typically water, or water based. Building-integrated energy systems are also presented. Here, renewable technologies are integrated into the fabric of buildings. Today solar architecture is undergoing a true revolution through the integration of renewable technologies into the fabric of buildings. Such systems are designed for heating, ventilation, thermal isolation, shading, electricity generation and lighting of buildings.

References

ASHRAE Fundamentals, (1997), Available at: http://www.filestube.com/ec305724730bb5c903e9,g/ash-rae-hand-books-1997-2000-p.html.

BSRIA (1999), 'Environmental Code of Practice for Buildings and Their Services', Available at: http://www.bsria.co.uk (Accessed: 07 August 2012).

Chemisana, D. (2011), 'Building Integrated Concentrating Photovoltaics: A review', in *Renewable and Sustainable Energy Reviews,* 15: 603-611.

Desbarats, G. (1980), 'Low energy Building Design Awards and Competition', *Minister,* Canada.

Djongyang, N. (2010), 'Contribution to the study of thermal comfort and coupled heat and mass transfer through building components in the Sub-Saharan Africa region', *PhD Thesis,* University of Yaounde I, Cameroon, 245 pages.

Dwyer, S.M. (2003), 'An Investigation of the Barriers that Exist for Building Integrated Renewables and their Implication for Sustainable Development', *Master Thesis*, University of Strathclyde, U.K.

D'Antoni, M. and Saro, O. (2012), 'Massive Solar-Thermal Collectors: A critical literature review', in *Renewable and Sustainable Energy Reviews*, 16: 3666-3679.

D&R International (2009), 'Buildings energy data book, buildings technologies program, energy efficiency and renewable energy', Report prepared for the U.S. Department of Energy, Silver Spring, Maryland.

Eicker, U. (2001), 'Solare Technologien für Gebäude', *B.G. Teubner*, Stuttgart/ Leipzig/ Wiesbaden.

Eicker, U. and Dalibard, A. (2011), 'Photovoltaic–thermal collectors for night radiative cooling of buildings', in *Solar Energy* 85: 1322-1335.

Fung, T.Y.Y. and Yang, H. (2008), 'Study on thermal performance of semi-transparent building-integrated photovoltaic glazings', in *Energy and Buildings*, 40: 341-350.

Harris, D.J. and Helwig, N. (2007), 'Solar Chimney and building ventilations', in *Applied Energy*, 84: 135-146.

Henemann, A. (2008), 'BIPV: built-in solar energy', in *Renewable Energy Focus*, 9(14): 16-9.

Hibshman, D. (1983), *Your Affordable Solar Home*, Sierra Club Books.

Hill, M. (2002), 'Friend of the Earth', in *Design Advice*, Issue 2.

Kachadorian, J. (2006), 'The Passive Solar House', *Chelsea Green Publishing Company*, White River Junction, Vermont, second edition.

Kapseu, C., Djongyang, N., Nkeng, G.E., Petsoko, M. and Ayuk Mbi Egbe, D. (2012), *Energies renouvelables en Afrique subsaharienne*, Harmattan, Cameroun.

Kelly, N.J. (1998), 'Towards a design environment for building-integrated energy systems: the integration of electrical power flow modelling with building simulation', *PhD Thesis*, University of Strathclyde.

Marta, J.N., Panao, O., Helder. and Gonçalves, J.P. (2011), 'Solar XXI building: Proof of concept or a concept to be proved?' in *Renewable Energy*, 36: 2703-2710.

Oliver, M. and Jackson, T. (2001), 'Energy and economic evaluation of building-integrated photovoltaics', in *Energy*, 26: 431-439.

Quesada, G., Rousse, D., Dutil, Y., Messaoud, B. and Hallé, S. (2012), 'A comprehensive review of solar façades: Opaque solar façades', in *Renewable and Sustainable Energy Reviews*, 16: 2820-2832.

Ralegaonkar, R.V. and Gupta, R. (2010), 'Review of intelligent building construction: A passive solar architecture approach', in *Renewable and Sustainable Energy Reviews*, 14: 2238-2242.

Robertson, K. and Athienitis, A. (2010), 'Solar Energy for Buildings', in *Canada Mortgage and Housing Corporation*.

Sadineni, S.B., Madala, S. and Boehm R.F. (2011), 'Passive building energy savings: A review of building envelope components', in *Renewable and Sustainable Energy Reviews*, 15: 3617-3631.

Schaeffer, R., Salem S.A., Frossard Pereira de Lucena, A., Soares M.C.B, Pinheiro P.N.L., Pereira F.F., Troccoli, A., Harrison, M. and Sadeck B.M. (2012), 'Energy sector vulnerability to climate change: A review', in *Energy*, 38: 1-12.

Shanthi, P.R., Sundarraja, M.C., Radhakrishnan, S. and Vijayalakshmi, L. (2012), 'Solar passive techniques in the vernacular buildings of coastal regions in Nagapattinam, Tamil Nadu-India: A qualitative and quantitative analysis', in *Energy and Buildings*, 49: 50-61.

Swanson, RM. (2000), 'The promise of concentrators', in *Progress in Photovoltaics Research and Applications*, 8: 93-111.

Wang, R.Z. and Zhai, X.Q. (2010), 'Development of solar thermal technologies in China', in *Energy,* 35: 4407-4416.

Website A: http://www.lightningcanyon.com/pages/passive-solar (Accessed: 10 August 2012).

Website B: http://www.omsolar.net/en/index.html (Accessed: 10 August 2012).

Website C: http://2.blogspot.com/Active+Solar+Heating (Accessed: 10 February 2012).

Zhai, X.Q., Song, Z.P. and Wang, R.Z. (2011), 'A review for the applications of solar chimneys in buildings', in *Renewable and Sustainable Energy Reviews,* 15: 3757-3767.

Chapter 3

Design Simulation and Realization of an Intelligent Charge Controller (ICC) for Accumulators in Renewable Energy Power Stations

Patrice D. Dongo[1], *Robert Tchitnga*[2], *Christian T. Tchapga*[3], *André R. Tchamda*[4], *Julius Kewir Tangka*[5] *&* *Anaclet Fomethe*[6]

1 Research Group on Experimental and Applied Physics for Sustainable Development, EAPhySuD, University of Dschang, Cameroon. Laboratoire des Energies Renouvelables, 'LER', Université de Dschang, B.P. 313 Dschang, Cameroon. Email: dongopat@yahoo.fr

2 Research Group on Experimental and Applied Physics for Sustainable Development, EAPhySuD, University of Dschang, Cameroon. Laboratoire des Energies Renouvelables, LER, Université de Dschang, B.P. 313 Dschang, Cameroon. Email: tchitnga@yahoo.fr

3 Research Group on Experimental and Applied Physics for Sustainable Development, EAPhySuD, University of Dschang, Cameroon; College of Technology, Department of Electrical and Electronic Engineering, University of Buea, Cameroon. Email: ctchito@yahoo.fr

4 Research Group on Experimental and Applied Physics for Sustainable Development, EAPhySuD, University of Dschang, Cameroon, Email: rodriguetchamda@yahoo.fr

5 Laboratoire des Energies Renouvelables, LER, Université de Dschang, B.P. 313 Dschang, Cameroun, Email:tangkajkfr@yahoo.fr

6 Laboratoire de Mécanique et de Modélisation des Systèmes, L2MS, Université de Dschang, B.P. 67 Dschang, Cameroun, Email: safomethe@yahoo.fr

1. Introduction

The importance of energy, for the worldwide development is known, be it fossil, fissile or renewable. Without energy, various branches of industries cannot perform operations related to: productions of goods, transport, health, habitat, supplies, evacuation of waste and recycling to list only these. However, new energies, in contrast to fossil energies, constitute a true alternative solution to limit pollution (see Website A).

Taking into consideration the intermittent character of the renewable sources of energy (wind, sun, water…), the development of storage techniques of the produced energy for later uses, like in case of insufficiency or absence of the sources becomes a requirement. The use of batteries for the purpose of storing electric power in chemical form is one of the found solutions. Batteries generally have a high cost, are fragile and are delicate in use. However, they constitute one of the most significant elements in the distribution's chain of the energy produced by power stations. The fragility of the batteries is amplified by the temperature of its ambient environment and the conditions, not always mastered, of charge and discharge. Messikh*et al.* (2008) considered the problem of ambient temperature in their work. They programmed their microcontroller to disconnect the batteries when the ambient temperature goes above the critical value of 40°C. With a view on optimal and rational use of the energy produced by power stations, we designed an intelligent charge controller that controls and maintains the ambient temperature in the optimal range of 15°C to 25°C as explained in the next section, where all the other integrated functionalities are presented.

The work is presented as follows: in Section 2, the theory behind the design of the proposed ICC is described. Section 3 introduces the steps followed in the programming while Section 4 is devoted to its architecture. In Section 5 the overall realization of the module is given, followed in Section 6 by the simulation and the discussion of the results. Section 7 is devoted to the conclusion.

2. General Description of the ICC

The ICC is a charge controller which is intended to charge batteries of nominal voltage 12 V, and capacity in the range from 40Ah to 120Ah. Two modes of functioning are foreseen: a normal quickly mode and a slow mode corresponding to the maintenance mode. For batteries with high capacity, only the slow charging mode can be ensured by the ICC since the needed current for their charge is lower than the fiftieth of their capacity. Once the source, be it of renewable, fossil or even of hybrid origin, is converted and adapted, the ICC admits at its entry a maximum continuous voltage of 100 Volts.

It ensures a long lifetime to batteries and the optimization of theirs output Gergaud (2002), taking advantage on its capacity to detect and control the temperature of the ambient air around the batteries. For in the case of the lead-acid battery on which the design of the ICC was oriented, the temperature's beach necessary for an optimal operation of the battery is ranging between 15°C and 25°C. It has been noted that an increase the ambient temperature of 10°C above the optimal upper limit taken at 25°C where batteries operate, can decrease by a factor of two (02) its life expectancy and, under the 15°C lower value, its capacity decreases (Gergaud, 2002; Midoun et A. Daoud., 2003).

The automation of the various processes in the ICC is commanded by a microcontroller PIC16F877: It ensures the management of all options that the module integrates, from where it's intelligent character. To achieve the extension, improvement and easier control of the functions of the ICC, following options are integrated to it:

- Automatic check of loads and remote communication with PC terminals;
- Protection against reverse current leakage and over-current;
- Protection against overload;

- Automatic detection and regulation of the ambient temperature around the battery by starting the heater or the fan;
- LCD signaling and LEDs indicators of the state of the source, of the battery, of the loads and of the ambient temperature around the battery;
- Sound signaling;
- Automatic disconnection and reconnection of the battery in the case of excessive charge respectively of major discharge;
- Luminous indication informing about the dysfunction of the operative apparatus;
- Optimization of the interface human-machine.

3. Algorithm and Flow Diagram of Operation of the ICC

Below is a legend of the symbols used for the presentation and description of the different functioning modes:

V_S : Voltage delivered by the source;

V_{ms} : Minimal voltage at the source that can still charge the battery (Threshold value);

V_b : Voltage across the electrodes of the battery;

V_{pb} : Voltage below which the battery is considered as absent;

E_{0nom} : Nominal voltage of the battery (equal to 12 V in the case of a lead-acid battery);

$E_{0\min}$: Minimal voltage below which it becomes critical to charge the battery (Threshold value: 10.8 V);

$E_{0\max}$: Maximum voltage above which the charged battery can be damaged (explosions danger).

Signal devices such as the LEDs (D3, D5, D7, D10, D12, D13, and D14) and the LCD make it possible to appreciate the behavior of the ICC. The algorithm of operation of the ICC is governed by the ten modes hereafter (See Figures 1 & 2):

- **Mode 0**: If low voltage at the source ($V_S < V_{ms} = 15V$) and low charge level of the battery considered as empty ($V_b < V_{pb} = 5V$), then disconnect the loads, maintain all LEDs off and update the posting on the LCD.
- **Mode 1**: If low voltage at the source ($V_S < V_{ms} = 15V$) and if the charge level of the battery limited by ($V_{pb} < V_b < E_{0\min} = 10.8V$), then disconnect the loads, switch on LEDs D7, D10 and update posting on the LCD.
- **Mode 2**: If low voltage at the source ($V_S < V_{ms} = 15V$) and if the charge level of the battery V_b is such that ($E_{0\min} < V_b < E_{0nom} = 12V$), then disconnect the loads, switch on LEDs D7, D10 and update the posting on the LCD.
- **Mode 3**: If low voltage at the source ($V_S < V_{ms} = 15V$) and if battery in economic mode, that is ($E_{0nom} < V_b < E_{0\max} = 14.8V$) then connect the load 1alone, switch on LEDs D3, D7, D12, D14 and update the posting on the LCD.
- **Mode 4**: If low voltage at the source ($V_S < V_{ms} = 15V$) and if the battery is fully charged ($E_{0\max} < V_b$) then connect the loads, activate alarm, switch on LEDs D3, D5, D12, D14 and update the posting on the LCD.
- **Mode 5**: If sufficient voltage is delivered by the source ($V_{ms} < V_S$) and the charge level of the battery empty ($V_b < V_{pb} = 5V$), connect the loads, switch on LEDs D3, D5, D7 and update the posting on the LCD.

Figure: 1 Complete Flow Diagram of the ICC

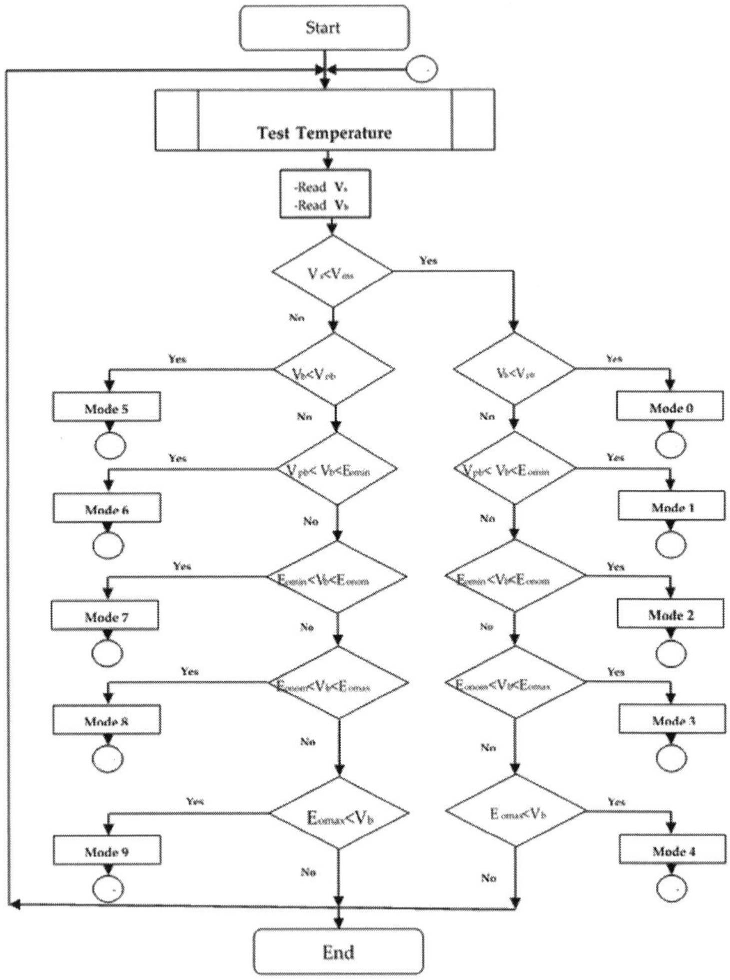

Figure 2: Flow Diagram of the Sub Program Test of Ambient Temperature around the battery

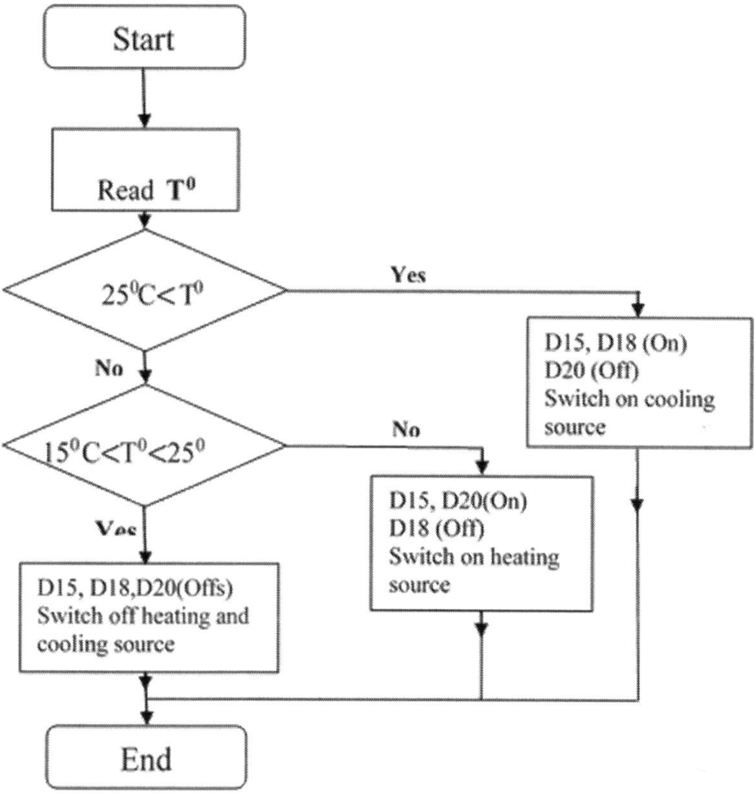

- **Mode 6**: If sufficient voltage is delivered by the source ($V_{ms} < V_S$) and the charge level of the battery too low ($V_{nb} < V_b < E_{0\min}$), then charge the battery, connect the loads, switch on LEDs D3, D5, D7, D10, D13 and update the posting on the LCD.

- **Mode 7**: If sufficient voltage is delivered by the source ($V_{ms} < V_S$) plus lightly discharged battery ($E_{0\min} < V_b < E_{0nom}$) then charge the battery, connect the loads and, switch on the LEDs D3, D5, D7, D10, D13 and update the posting on the LCD.

- **Mode 8**: If sufficient voltage is delivered by the source ($V_{ms} < V_S$) and battery charged ($E_{0nom} < V_b < E_{0\max}$), then continue charging the battery, connect the loads, switch on D3, D5, D7, D13 and D14, and update the posting on the LCD.

- **Mode 9**: If sufficient voltage is delivered by the source ($V_{ms} < V_S$) and full charged battery ($E_{0\max} < V_b$) then disconnect the battery, connect the loads, activate alarm, switch on LEDs D3, D5, D14 and to update the posting on the LCD.

4. Architecture and Presentation of Various Blocks of the ICC

The synoptic diagram of the ICC given by Figure 3 consists of six principal blocks of which the intern constitutions are as follows:

Figure 3: Synoptic Diagram of the ICC

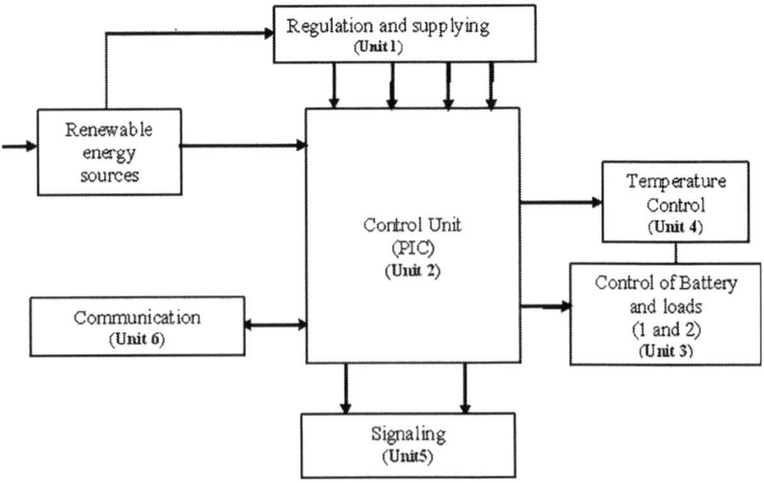

Unit 1: Regulation and Supplying (Zaamta et al, 2009; Power Electronics, 2006)

This unit provides a stable DC voltage of 15V necessary for charging the battery and for supplying the following Blocks. It is essentially made of a zener follower, which is a linear regulator, constituted of zener diodes, assembled in reverse and supported by a power transistor. This operative apparatus is able to deliver a very significant current at its output. Rectifying diodes are assembled in series with the zener ones in order to compensate the voltage drop imposed by the base-emitter junction of the power transistor.

Unit 2: Control Unit (James and Dunlop, 1997; Bigonoff, 2003; Microchip, 1999; Balogh, 1997; Mayeux, 2002)

Unit 2 is supported by the microcontroller PIC (Programmable Controller Interface) 16F877. The PIC measures the voltage served to the battery by the DC/DC converter. It then compares it with the stored threshold value. According to the result, it can generate through a Pulse Width Modulation (PWM) command and at a different mode, the necessary signal for charging the battery and supplying the loads. It is thanks to the PIC that the programming and the automatic management of tasks assigned to the controller are well conducted. The choice of PIC 16F877 was guided by its high number of pins (40) which is helpful when many processes are to be commanded, its capability to convert analog to digital signal, its high number of entries and its huge memory capacity (8192 words, 14 bits) for programs.

Unit 3: Control of Battery and Loads

This unit constitutes the power part of the system. It is essentially made of relays which command the connection and disconnection of loads, respectively of the battery. We assumed that two loads must be managed by the controller. The first would be a principal load 1, for example a public clock, a refrigerator for the conservation of vaccines and drugs in a hospital etc., having priority on the secondary load 2. Depending on the characteristics of the source and the current requested by the loads, the control unit offers to the ICC a capacity of managing several loads.

Unit 4: Control Temperature

This unit aims at keeping the battery in an operating optimal temperature interval. With help of the linear temperature sensor LM 35, which has the advantage of having a good precision Power Electronics (2006), the temperature is detected in the vicinities of the battery. According to its value, an automatic regulation is set, making it possible to optimize the lifetime of the battery. By the mediation of the control unit and the interface of power, the ICC integrates functions ensuring regulation.

Unit 5: Signaling

This unit makes the communication between the operator and the machine possible by bill-posting on the LCD (Liquid Crystal Display) and by setting on or off LEDs (Light Emitting Diodes). Bill-poster LCD used here is a recycled NOKIA 3310 trademark display. The bill-posted information concerns the status of the source, of the battery, of the loads and of the sensor for temperature, as well as their securities. The LEDs inform equally on the state of the control's parameters. The voltage supplying the display of type NOKIA 3310 varies between 3 Volts and 3.3 Volts. It consumes a current of 0.2 mA to 0.8 mA. It carries 8 pins, for a size of 40x38 mm^2.

Unit 6: Communication

The ICC device is connected to a computer via a serial port which ensures a bidirectional communication between both. Using this means, a control of evolution of the parameters relating to the system can be achieved via the computer:

voltage of the source, voltage across the battery and ambient temperature. To achieve the voltage compatibility between +12V at the serial port of the computer and the 0V to +5V of the logic TTL of the PIC, an integrated circuit MAX 232 assisted by the capacitors C7 to C9 is inserted into the diagram (See Figure 4). The two capacitors enable the pumping of the voltage from the +5V at the PIC to +12V needed by the power elements of the circuit (Garot, 1985; Power Electronics, 2006; Website B; Website C).

Figure 4: Symbolic Representation of IC MAX 232

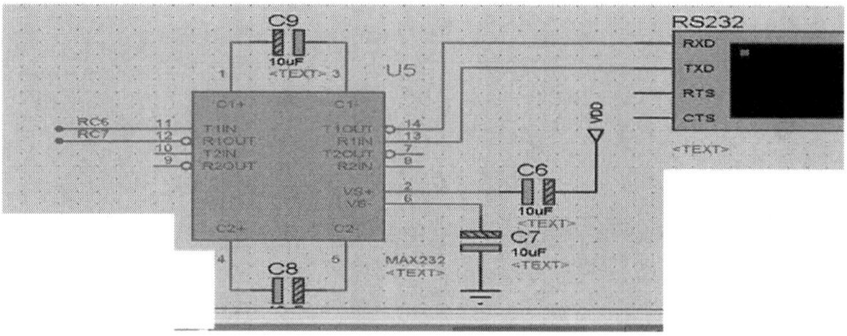

5. Design and Realization of the ICC

Using the electronic simulator ISIS version 7.5, the complete diagram of the ICC is drawn below (Figure 5), with the appropriate components.

Figure 5: Electrical Diagram of the ICC

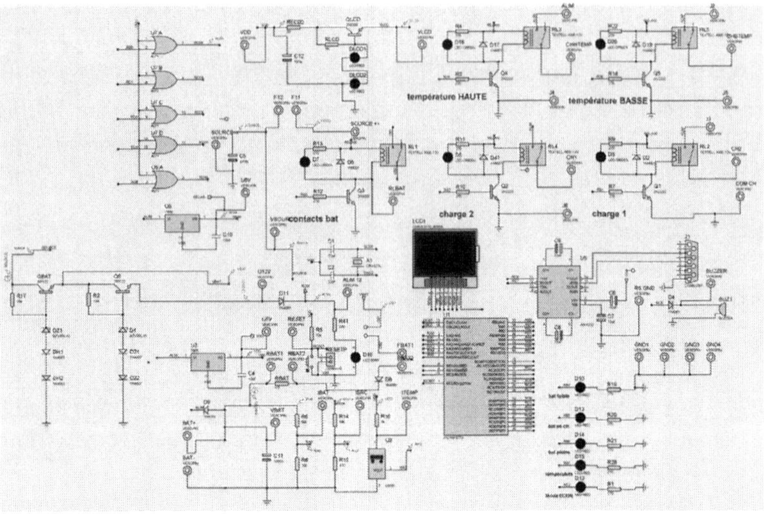

The physical model obtained after the real implementation of the components is presented on Figure 6.

Figure 6: Picture of the Implemented ICC

6. Simulations of Results

The results of simulation obtained are located in Table 1. The LCD informs on the behavior of parameters at the power station. The LEDs from D3 to D15 in Table 1 characterize precise states of functioning of the device and complete the information summarize on the LCD. Table 1 illustrates the ten modes of functioning recorded for the charge controller. The results are more explicit on the LCD screen.

The results generated by the simulator and gathered in this table are in perfect agreement with those provided by the flow chart. The LCD posts throughout this simulation the mentioned "Normal temp" message, confirming that the battery operates in the temperature range between 15°C and 25°C which is the range for normal temperature. This confirms that this range of temperature is necessary for an optimal operation of the battery.

a) **For** $15°C < T° < 25°C$

Table 1: Results of both simulation and manipulation for $15°C < T° < 25°C$

Signalling / Mode	D3 C_1	D5 C_2	D7 batt	D10 Low Batt	D12 Batt. in Econ Mode	D13 Batt. in charge	D14 Batt. full	D15 (T°)	LCD screen (Simulation result)	LCD screen (experimental result)
Mode 0	Off	Off	Off	Off	Off	Off	Off	Off	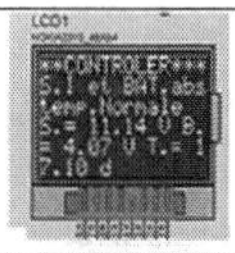	
Mode 1	Off	Off	On	On	Off	Off	Off	Off	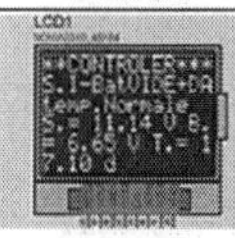	
Mode 2	Off	Off	On	On	Off	Off	Off	Off		
Mode 3	On	Off	On	Off	On	Off	On	Off		

Mode 4	On	On	Off	Off	On	Off	On	Off	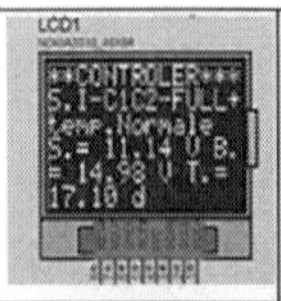	
Mode 5	On	On	On	Off	Off	Off	Off	Off		
Mode 6	On	On	On	On	Off	On	Off	Off		
Mode 7	On	On	On	On	Off	On	Off	Off	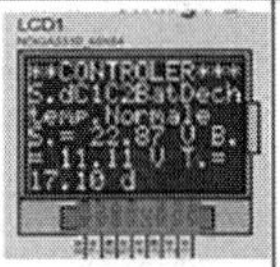	
Mode 8	On	On	On	Off	Off	On	On	Off	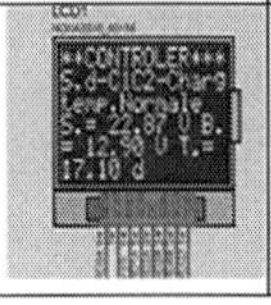	
Mode 9	On	On	Off	Off	Off	Off	On	Off	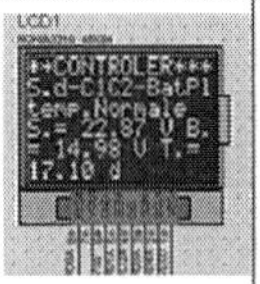	

Discussion of the Results (Case of mode 7)

According to the information displayed on the screen both experimentally and by simulation (see Table 1) we can notice that the results are perfectly correlated. For the simulation result we have the following information: S.=22.87V, B.=11.11V and T.=17.10d. "S." stands for the source, "B." for the battery and "T." for the temperature respectively. We can read on the simulated display "S.dC1C2BatDech" and "temp. Normale" meaning that the source is available (S.d: Source disponible *in French*) because the voltage is above 15V; loads are connected (C1C2: load 1 and 2 *from French Charges*) because the energy at the source is available. The battery is discharged, but not deeply (BatDech: Batterie déchargée), since the voltage across the battery is less than 12V. Meanwhile the ambient temperature is normal that is between 15 °C and 25 °C. For the experimental result, similar information appears on the Nokia Display.

b) For $T° > 25°C$

The results obtained for each mode evaluated previously are identical. One observes in addition to the previous that the LEDs D15 (critical Alarm temperature) and D18 (Activation of the ventilator) are on. It means that the ambient temperature around the battery is above the threshold value of 25°C, hence no optimal condition for the battery. The controller puts on the fan automatically to control the ambient temperature. At the same time, one can read on the display: "T. Critik/chaud", which means critical hot temperature and starting of the fan (Figure 7).

Figure 7: Presentation of bill-poster LCD for $T° > 25°C$
c) For $T° < 15°C$

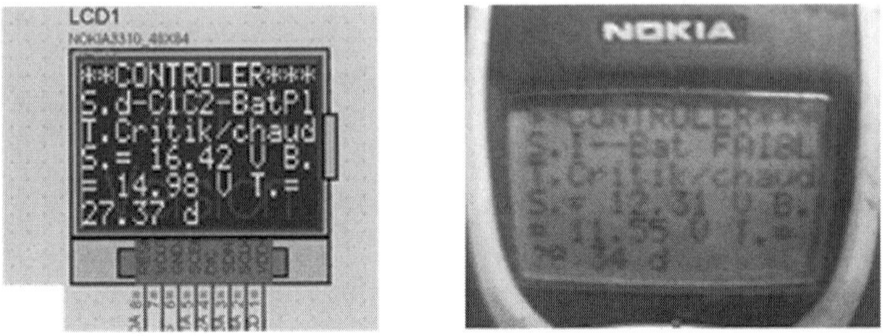

The results obtained for this range of temperature for each mode evaluated previously are identical. One observes in addition that, the LEDs D15 (critical Alarm temperature) and D20 (Activation of the heating) are on. It reveals that the temperature around the battery is low for its optimal operation. The controller puts the heater automatically on to bring back the temperature to the necessary normal range.

In this range, the LCD displays "T. Critik/froid", which means critical cold temperature and necessity to heat (Figure 8).

Figure 8: Presentation of Bill-Poster LCD $T° < 15°C$

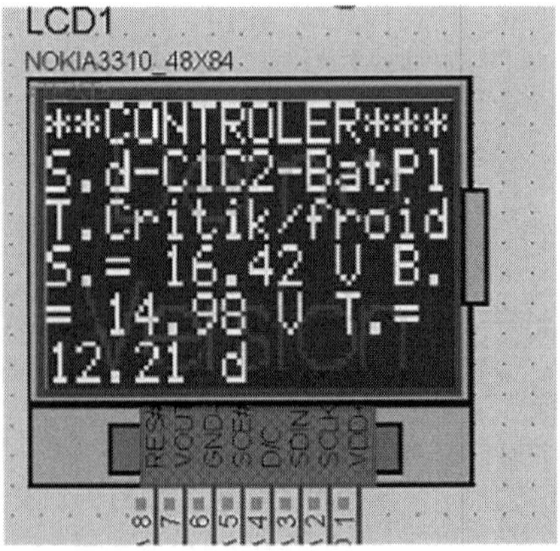

Figure 9: Experimental Circuit in the Functioning State

7. Findings

The research has found that the designed and realized intelligent charge controller ICC is capable, among other things, to manage in an efficient manner the cycles of charge and discharge of batteries in power stations. It also allows managing in a rational and optimal way the energy produced by the power stations of renewable energies. The extra feature of considering the ambient temperature around the battery as an important parameter for the lifetime of batteries makes it more original compared with many other existing charge controllers.

The results obtained fit suitably with the operating modes defined, so the ICC satisfies the expected specifications assigned to its design. The electronic device ICC is also capable, considering the harmonious choice of its components and their dimensioning, to meet multiple needs; because the design of the ICC rests on a programmable logic, which makes it very flexible. All things considered, the ICC will make it possible to contribute in a significant way to the efficient use of batteries and the energy management, mainly of new energies for a popularization and a better use.

References

Balogh, L. (1997), 'Implementing Multi-State Charge Algorithm with the UC3909 Switch mode Lead-Acid Battery Charger Controller', *U-155 Application Note,* Unitrode Applications Handbook, pp.3-488 – 3-516.

Bigono, FF. (2003), 'La Programmation des PIC par Bigonoff', Seconde partie (PIC16F876-16F877), 7ème Révision.

Garot, P. (1985), 'Manipulations et travaux pratiques d'électronique'' Edition DESFORGES, Paris.

Gergaud, O. (2002), 'Modélisation énergétique et optimisation économique d'un système de production éolienet photovoltaïque couplé au réseau et associé à un accumulateur'', thèse de Doctorat soutenue à l'école normale de CACHAN.

James, P. and Dunlop, P.E. (1997), 'Batteries and Charge Control in Stand-Alone Photovoltaic Systems', *Fundamentals and Application,* Florida Solar Energy Centre, January 15.

Mayeux, P. (2002), 'Apprendre la Programmation des PIC par l'Expérimentation et la Simulation', plus un CDROOM, ETSF 2ème edition.

Messikh, L.S., Chikhi, F. and Chikhi et T. Chergui. (2008), 'Mise au point d'un régulateur de charge / décharge de batterie avec seuils adaptatifs de tension pour les applications photovoltaïques', *Revue des Energies Renouvelables,* 11 (2) : 281 – 290.

Microchip (1999), 'Datasheet 'PIC 16F87x: 28/40 pin CMOS Flash Microcontrollers', Microchip Technology Inc. DS30292B.

Midoun et A. Daoud, (2003), 'Commande Floue de la Charge d'une Batterie dans une Installation Photovoltaïque', *Revue des Energies Renouvelables 'ICPWE'*, pp. 67 – 72.

Power Electronics (2006), EE362L, Power Electronics, 'DC-DC Buck Converter', Version, February 28.

Website A: http://www.sonelec-musique.com/électronique (Accessed: 23 March 2012).

Website B: http://www.datasheet.com (Accessed: 25 January 2012).

Website C: http://www.Iepf.org (Accessed: 10 January 2012).

Zaamta, S. et A. Dib. (2009), 'Réalisation d'un régulateur solaire à base de microcontrôleur pour le contrôle del'état de charge et la protection des accumulateurs', in *Revue des Energies Renouvelables,* 12 (1): 137 –147.

Chapter 4

Methodology of North-South Technology Transfer: Application to the Case of a Foodstuff Drier

Yacoub I. Halawlaw[1] *& Mahamoud Y. Khaya*[2]

1. Introduction

In the countries of the South, many farmers lack both the means and knowledge to apply various technologies necessary to the development of their exploitations; technologies which they badly understand and have difficulties to integrate in their usual practices.

The process industries in the developing countries did not adopt yet a system of hygiene and management of the environment, in conformity with the international standards, very often because they are under-informed on the standards and relevant codes of good practice imposed by the importers.

1 Department of Physics, Faculty of Pure and Applied Science, University of N'Djamena, N'Djamena, Chad. Email: yacoubih@gmail.com

2 Institut Universitaire des Sciences et Techniques d'ABECHE, Chad. Email: mykhayal@yahoo.fr

In addition, in the developing countries, there exist several institutions of quality control but, in the majority of cases, they show an obvious lack of coordination and co-operation.

The adequate quality control requires a checking of the results at the laboratory. In the developing countries, there are not enough laboratories, and those existing do not have the capacity necessary to carry out certain tests, which disqualify them in the eyes of Western importers.

Theoretically, all the countries of WTO are invited to take part in the development of the international standards. But in reality, the countries of the South are marginalized when working out the standards of the Codex.

In addition, the techniques are very limited. Industry is embryonic. Put aside the oil producing countries, the infrastructures are quasi-non-existent. The trade balance is always adverse. Thus, certain African states are dependent on the external assistance, even to pay their civil servants. Let us add to that bad management, corruption and the diversions of public money. The delinquency and the unemployment of the youth are amplifying. Can one consider under these conditions economic development? Or should they on the contrary, be left with their sad fate? What is to be done to help these countries to leave the stagnation in which they are plunged? The actors of North think that it is possible to improve the situation thanks to the transfer of technology operations.

But the problem with the current technology transfers, lies in the fact that one was always satisfied to transfer from the North equipment, without taking into account the environment of recipient country, with the risk that, this equipment does not function, or function during a certain time and fall in disuse in spite of its proven performances, under the conditions of industrialized countries.

A need for suitable methods which can put all the chances of success on the side of the promoters of such operations emerges. Our guiding principle consists of thinking the technology transfer as a procedure of innovation. Thus defined, the technology transfer can be analyzed like a procedure of innovation and apply to it the theories of innovation.

The objective is not to emphasise the technical details of designing a foodstuff drier; it is more to demonstrate how to organize these details to be able to propose a drier that may work in a southern environment.

2. The Classical Approach

The international transfers of technology are not exclusively a phenomenon of the 21st century. One of the first technology transfers known in history took place at the XIIth century of our era under the dynasty of Abbasids during the reign of Caliph Haroun Al Rachid (Amin, 1974). The scientific and literary activity was so intense that paper periodically was lacking. Stocks were quickly exhausted, and it was necessary to import the paper from China where paper was manufactured. The sovereign decided to have no more recourse to the importation of paper from China, because the procedure was not only expensive, but the voyages were very long and perilous. Then the caliph brought the specialists of paper to Baghdad where they taught to local volunteers the paper mill trade. It was one of the very first successful technology transfers. The key to success, in this precise case, was the interaction which took place between the actors implied in the transfer operation. The Chinese specialists taught their methods to local disciples (it is what we call interaction); the latter with the assistance of the first applied to the ground the knowledge received by using local materials and local competences. Thus, the technology transfer, who took place more than eight centuries earlier, was conceived like a step of innovation. One recreated the conditions of the manufacture of paper by taking into account the new environment stimulating thus, local creativity. Today this method, which proved reliable in the past, was completely forgotten not to say completely ignored. Most technology transfers follow a rigid and unsuited scheme, which consists in moving technical objects having proven reliable in a given socio-cultural environment, towards another without taking into account the characteristics of the environment of reception.

The reasons usually advanced, to explain the failure of certain transfer of technology operations, are rather superficial. They concern "the opening" or the "closing" of cultures to progress (Desmorieux *et al.*, 2001). Noted failures here and there (Akrich *et al.*, 1988) have a rather profound cause: it is at the level of the conceptual approach in the technology transfers.

3. Explaining Technology Transfers

In industrial circles, one understands by technology transfer "the passage from the laboratory knowledge to industrial know-how" (Durand, 1994)

This passage of the technological knowledge to the industrial practice is done by stages: One passes from a state of knowledge to the invention and from invention to the innovation. The innovation is the first use of invented technology. The process which leads from technology to the technique, successively implements the fundamental research, the applied research, the development of the prototypes and finally industrialization in the form of series production of articles or consumer goods.

For Rouach and Klatzman (1993), any diffusion of new technique is a technology transfer. Or each time that a knowledgeable person invented a new instrument and that this instrument was imitated and used by others, a process that can be defined as the technology transfer takes place.

According to UN (Rouach and Klatzman, 1993), the technology transfer is the transfer of knowledge necessary to manufacture a product, to the application of a process or the performance of a service and does not extend to the transactions comprising the simple sale or simple hiring of goods.

We propose to define the technology transfer as a space-time movement of technical objects, knowledge, know-how, and competences in the field of technology with an aim and impact of social transformation and economic development.

According to this definition, one does not only move technical objects from a given environment towards another, but also the production methods of these objects, stimulating thus in the process, local creativity.

4. The Moroccan CNTTA Approach

Furthermore, several international organizations devote the development aid to the technology transfer perceived as an effective means to fight against underdevelopment. Figure 1 illustrates the Moroccan CNTTA Approach.

Figure1: The Moroccan CNTTA scheme

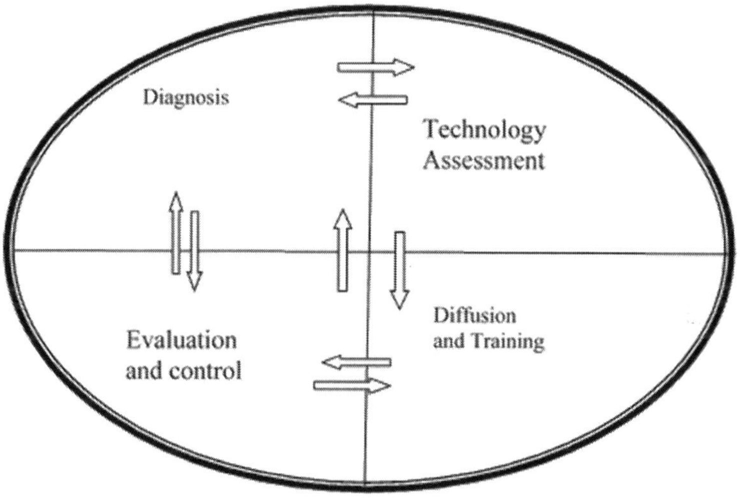

- The diagnosis is an activity which consists in analyzing the current situation in order to identify the constraints, to treat them on a hierarchical basis and finally to seek the suitable solutions in collaboration with the farmer. The methods used are generally based on investigations of ground and/or work into documents. If the solutions exist, one passes to the following phase, if not a stage of research and experimentation is necessary.

- Technology assessment makes it possible to be ensured of the adaptability of the solutions available and suggested, in the technical and socio-economic conditions of the farm. The experimentation, in the farms, and its follow-up, constitute the adapted tool for this activity. The objective of this experimentation is to prove the feasibility of the technology suggested under the real operating conditions.
- Evaluation and control: The ultimate test of the quality of a technological or socio-economic innovation is its rate of adoption by the target group. The objective of this phase is the follow-up of the adoption of technology, the analysis and the interpretation of the results obtained by the farmers as well as the detection of problems having emerged, at the end of the implementation on a large scale of the innovation suggested.
- Diffusion relates to technologies which passed the test and brought to be diffused on a large scale. The actions of this phase are based on the training of the promoters and the diffusion of technology. The tools used are information, formation and demonstration.

This method which proceeds by stage with the possibility of checking the results at each phase and making the necessary corrections is interesting but it supposes that technology is diffused mainly due to its intrinsic qualities, underestimating the use that would make the target populations of them, but a new use of a technical object guarantees its appropriation by the social group for which it is intended (the eloquent example of this fact is the car in Africa and more recently the mobile telephone).

5. The SEC Approach of FAO

Moreover, the experts of FAO (Adhikarya *et al.*, 1994) worked out a methodology of technology transfer entitled SEC (Strategic Extension Campaign). This is illustrated by Figure 2.

Figure 2: The FAO SEC Scheme

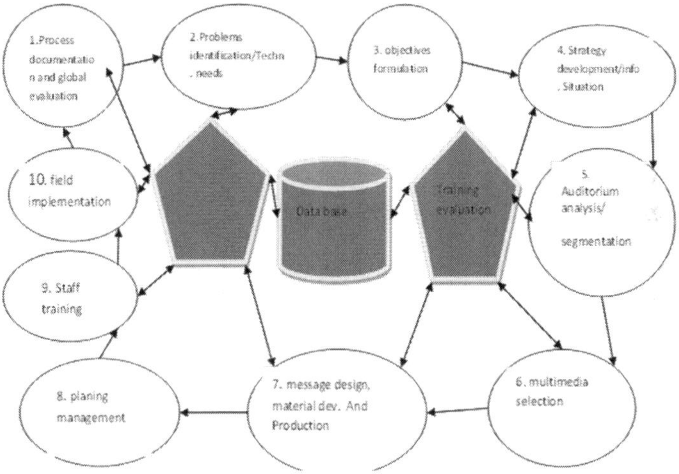

This approach has many merits, but it considers that the target populations are passive containers which we must fill of technological innovations and it does not associate them at the phase of design of the projects.

6. Limits of the Classical Approach

The general scheme remains the same: The countries of the South are containers which it is necessary to fill by artefacts and knowledge. This channel is subjected to the action of noises of any kind. We understand by noises the transformations which can undergo the object to be adapted to the local conditions and the measures of accompaniment required by this adaptation. Such a scheme leads as well as possible to the recovery of the object transferred and worst to the loss of this object which falls in disuse as soon as the promoters of the step withdraw themselves.

The reasons advanced until now to explain these failures are not very convincing since they are founded on judgments formulated according to Western criteria: lack of maintenance, the African spirit "nonchalant" (In a television broadcast one shows Africans which use computers as simple kitchen tables). One to some extent tries to show that the underdevelopment is an inevitable calamity due to the culture of the people of the developing countries.

7. STOU Approach

Figure 3: The STOU Scheme

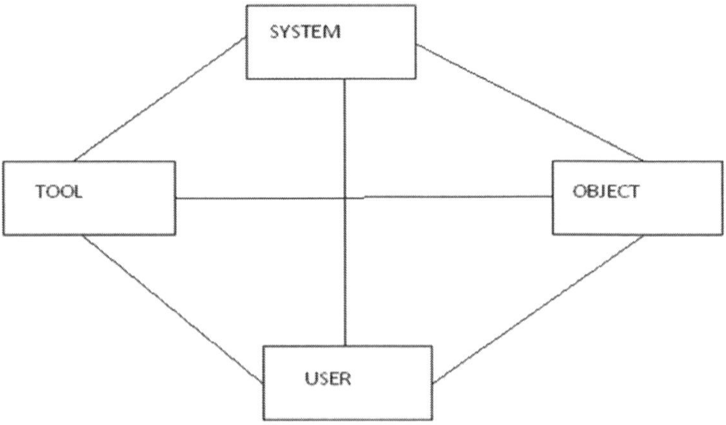

Figure 3 shows the STOU model (S-System T-Tool, O-Object, U-User), which describes the relations created by and around an instrument, this one being applied to an object. The analysis is founded on the quartet represented on the figure above

- S- The surrounding system (for example: the support of the drier, the price of Spirulina, the use of the dried Spirulina)
- T- the tool (the drier)
- O- the object (paste of Spirulina)

- U- the user (of the drier)

The basic relation consists the bond between the object and the user (O-U);the relation O-U (Spirulina-User) creates the need materialized by the bond (U-T-O); what can be formulated like the user (U) needs the drier (T) to dry Spirulina (O). This first analysis which gives the triangle U-T-O is not enough to define the real conditions of use since each element is in relation with a broader environment which is social, physical, and economic. More generally the relation of the user to the system passes by the sequence of all the relations (U-T-O-S). Thus, the technical objects designed in North are integrated in a system and found a certain balance within quadruplet STOU.

The decision to move them just as they are, in a new socio-economic and cultural environment, means transposing STOU into S'TOU' brutally. What means putting equality between S and S'; U and U' and this is not the case. From there, one sees easily the many failures noted in the technology transfer operations.

It should be stressed that even T in the S' system is no more T but become T' and even O in the S' system become O'.

This model explains well the causes of the noted failures, but it does not propose any step which could lead to success in the projects of technology transfer.

8. Innovation in the Processes of Technology Transfers

We showed previously that the approach, until now used in the technology transfers, towards the developing countries, led to failures in most of the cases. The reason was that, this approach is founded on a mechanist theory of the technology transfer. It is restricted to move technical objects designed and manufactured out of context and thus not adapted to the local conditions (Yacoub, 1998, 1999, 2010; Yacoub *et al*, 2006).

To avoid or circumvent these difficulties, we advance a fundamental assumption which we will adopt all along our step: the technology transfers must be thought as a step of innovation. Thus, the technology transfers must stimulate local creativity. The future users of the equipment which one wants to transfer must be in one way or another associated to the design of the aforesaid equipment and their manufacture must (at least partly) be carried out on the spot. One thus eliminates all the problems due to maintenance and the supply in spare parts which are one of the principal reasons usually advanced to explain the failure of certain technology transfers. Based on the fundamental principle stated higher, we propose a new methodology of North-South technology transfer (METTENS).

9. Proposed Methodology for North-South Technology Transfer

Figure 4 presents the proposed methodology for the North-South Technology transfer. To synthesize, and to make a successful North-South technology transfer, it is necessary initially:

- To study sufficiently the socio-economic and cultural context of transfer places (Northern and Southern)
- To make a serious analysis of the needs, that should yield an acceptable evaluation of the demand.
- To carry out a market research, in order to obtain acceptable assumptions on the state of the market
- To form a primary network gathering the local actors.
- Within the framework of the primary network, create the intermediate objects necessary to the interaction of the members of this network and thus of the interlocking of its dynamics.
- The primary network being opened, it must be extended gradually to other members.

- The network thus widened must open to other networks, which would lead to the setting-up of new intermediate objects qualitatively and quantitatively new.
- Formation of a super network gathering the various networks (It is the mother of all networks).
- This last Super network has to concretely carry out the objectives of the project of transfer.

We understand by network, the unit which gathers humans, the artifacts and the products to which these artifacts apply. The actor is for us human, an artifact or a product to which this artifact applies. The order is not obligatorily necessary: it is possible to start with the formation of the networks and then to carry out the analysis of the needs or conversely. Our scheme does not exclude the return tickets and the modifications during action.

Figure 4: The Proposed Methodology Scheme

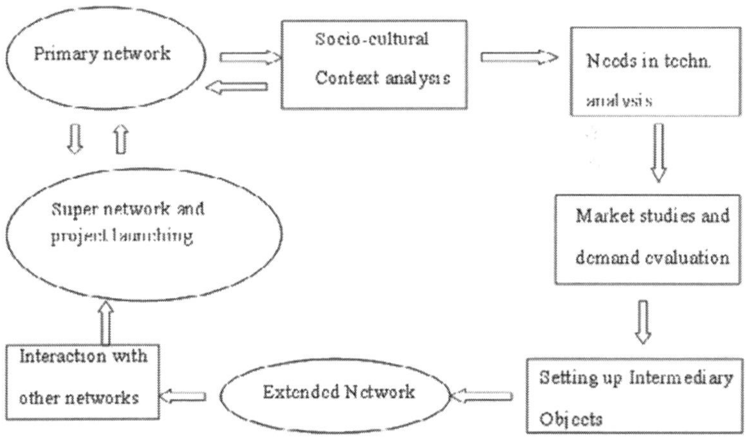

The study of the socio-cultural context of the actors taking part in the transfer of technology operation fulfils the requirement which consists in probing the ground, on the two sides (North and South) in order to provide the foundations of the analysis of the needs in technology. This stage is essential, since it must update the socio-cultural characteristics of the various actors, in order to allow the promoters, to seek to transfer only the technical objects likely to answer, effectively the expectations of the target populations, without running up against their spiritual values appreciably. The technology transfer is a step of innovation which meets multiple barriers at the entry. Approaches based on the local creativity and the local experts are likely to make a success, because such approaches build on the continuity of the knowledge to create innovative action.

The analysis of the needs is the following stage. But what kind of analysis we must perform? Which needs? No step of technology transfer was done without an analysis of the requirements in technology. However, these analyses were based on western-centric assumptions on what is necessary to the populations of the countries of the South and on what is not. But the experiment shows that such considerations are generally not justified.

It is advisable to associate the recipients, at the phase of design. An analysis of the needs realized jointly with the local populations, is more realistic and would be adapted better to the local conditions.

During this stage, a group is necessarily formed, charged to formalize all the requests of the recipients in technical terms easy to interpret by Survey Office. It is this starting group which forms the primary network, essentially made up of local elements. The role of this primary network consists in setting up the intermediate objects, which engage its dynamics. These objects would be the privileged place of the interaction of the actors of the network. The network is by definition open. What implies the opening to other actors or networks; that leads on a side, to the setting-up of new intermediate objects along with the widening of the network on the other side.

Regarding the primary network, it is necessary to raise the following questions: who must be member of this network? How to carry out the formation of this network? Which are the intermediate objects which must set up the primary network?

Intermediate Objects

They are in fact objects produced or used during the process of design, traces and supports of the action to conceive, in relation to the tools, the procedures and the actors. Their empirical contents can remain fuzzy and to be given only individually according to what is presented on each ground. Their interest is to enable us to lay out on a concept with the same level of generality as that of organization and to put questions relating to the study of the processes without disregarding their concrete way and their instrumentation, their contents to function (Jeantet, 1998).

- The feasibility study of project consists in justifying to public or private organizations and financial organizations, the profitability and the economic effectiveness of the project. It is composed of a general presentation of the project, history of the question, program of development considered, technical studies of the sites of the project as well as the economic study of its realization. The essential goal is to obtain the green light for the release of the project and the agreement of financing.
- Analysis of the needs: this stage is essential, and it is recommended to devote a rather sufficient time to it, during the process of transfer and design of the equipment to be transferred. Indeed, one of the principal problems which one encounters in the developing countries, milked with the lack of reliable information and the difficulty of communication. These problems come from the successive adaptations of the equipment to the potential users, without possibility to put in question the basic concepts, yielding therefore technical complexities and an

increase in the costs. It is also the stage which requires the maximum of competences (sale, economy, ergonomics, manufacture, processes, products, sociology, animation and marketing). It is necessary to note, at this stage, that the encountered problems are specific to a given place or they can be met everywhere; if it is the case, it is necessary to get information about the means that use others to solve them. It is then advisable to specify which are the users aimed for the use of the equipment which one wants to transfer and to associate them to the preliminary phase of the transfer operation. Furthermore, it is necessary to ask the question to know which the awaited functions of the equipment are. Not only technical functions but also functions of use and functions of regard. Then, it will be a question of classifying these functions according to their importance. The observations necessary to the analysis of the needs will also be necessary to the ergonomic analysis of the behaviour of users of equipment existing in nearby fields. A market research is essential although the means are limited in the developing countries compared to the country of North.

- Functional Task Book (FTB): This FTB presents the awaited functions of the future equipment, their criteria of appreciations, their levels as well as the constraints of the project, without the fork of the cost of acquisition. One should not forget competition in the field. Displacements in zone of use of the material will also make it possible to identify the modes of diffusion and maintenance of equipment similar to those considered.
- The Functional Task Book: the standard X50-151 (Delafollie, 1996) gives the following orientations: The Functional Task Book is a document by which the applicant expresses his requirement (or what it is charged to translate) in terms of functions of services and constraints. For each one of them, criteria of appreciation and their levels are defined. Each one of these levels must be supplied with flexibility. In our case, FTB is a stage necessary before the development of the

plans and the setting-up of the prototypes of the equipment concerned. It consists in specifying the functions which the equipment must fill. FTB does not propose concrete technical solutions. It is rather used to guide the choice of the designers. The establishment of FTB implies that a study made it possible to determine with precision the user's needs. The objectives are to obtain an answer to the proposed product ready to fulfil the awaited functions, under the conditions envisaged, at minimal cost. For this goal, the FTB expresses only requirements of result and, in theory, any requirement of means. The criterion of appreciation (of qualitative nature) is accompanied by a scale making it possible to locate its level (of quantitative nature). The FTB is composed of:

- A general presentation of the problem: that consists of the orientation of action VA and the information retrieval.
- The functional expression of the need is the essential phase of the FTB
- The call to alternatives, if it is necessary
- The framework of answer: it relates with the research of ideas and the ways of solutions including the monitoring and the evaluation of the solutions.
- At the end of the setting-up of the FTB, one proceeds to the choice of concrete technical solutions. These solutions are translated in the form of plans and of technological drawings which will be used as a basis for the development of the prototypes. After tests on the ground, prototypes will open the way to the series production of the equipment considered.
- The definition of the artifact: it is the last phase of thinking before complete materialization of the equipment. It is now a question of defining, in a precise way, the future equipment, the manufacturing processes and the provisioning. The principal tools used are the instruments of design engineering (worksheets, catalogues of components, drawing boards or CAD) and those which allow taking into account the provisioning,

the manufacture and the maintenance of the future equipment. It is essential to define the components of the product according to the constraints related to the manufacturing process, with the availability of materials and the functions to be filled. Then, it will be necessary to do an inventory of local know-how and also specify the possibilities of local maintenance. At the end of this stage, the Task Book of the Product (TBP) will have to be written in the form of note and industrial development programs of calculation and indications of construction.

- The construction of the prototype: here, it is the turn of the local equipment suppliers, associated at the beginning of the project, to make the financial decision to invest in the prototype, then to make all the financial efforts for the launching of the new products. The contribution of the foreign backers will come to supplement the effort of the local equipment suppliers and does not have to be the single source of financing, to avoid opportunist behaviours which consist in selling at the foreign financial organizations the equipment at prices three or four times superiors at market prices. The prototype being built, one passes then to his validation before the launching of the production on a large scale of the product.

A socio-technical network can be represented as shown in Figure 5:

Figure 5: Representation of a Socio-Technical Network

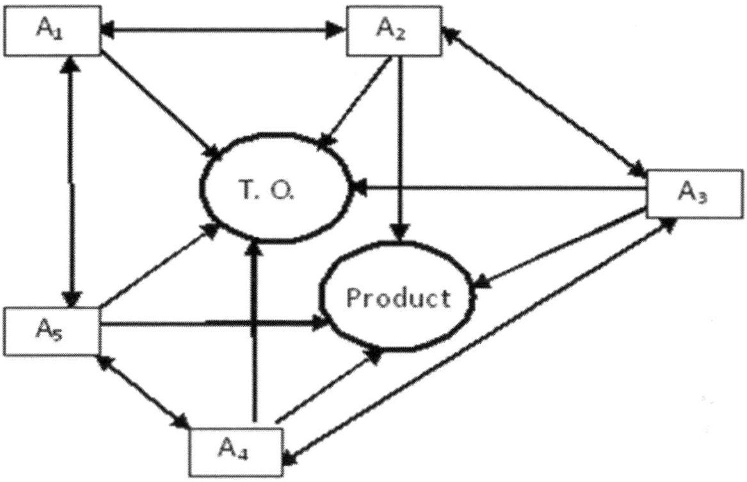

Caption: With A_i is an actor engaged in the project, TO is the technical object (drier for example) applied to the product (Spirulina for example), the arrows with one direction indicates the interest of the actor for the TO and/or the product to which the TO is applied and the arrows with double direction indicate the various types of interactions (socio-technical relations) between the actors taking part in the step.

In Table 1, the possible interactions between the actors participating in the step is provided.

Table 1: Matrix of Interaction of the Actors by Means of the Intermediate Objects

O.I/Actor	1	2	3	4	5	6	7
Humanitarian organizations	C	C	V	V	V	V	V
Political authorities	V	V	V	V	V	V	V
Survey Office	V	V	C	C	C	C	V
Researchers	C	C	C	V	C	V	V
Financial organizations	V	V	V	V	V	V	V
Tradesmen	V	V	V	V	V	V	V
Manufacturers	V	V	V	V	V	V	C
Maintenance workers	V	V	V	V	V	V	V
Consumers	C	C	V	V	V	V	V

Note on Captions: 1... Analysis of needs, 2... FTB, 3... Principles of solutions, 4... Overall drawings, 5... Digital Simulation, 6... Detailed drawings, 7... Prototype, C means likely to create the object; V means validates the object.

The concept of innovation that we attach to the concept of technology transfer is not less complex. It is also function of the point of view of the actor considered. Also, it is advisable to treat these two questions in a global way, by considering all the variables. But it is a known fact of the specialists, that in the case of innovation and thus of technology transfer, the number of the variables to consider is variable.

We considered the technology transfers, from the rather restricted point of view, the development of a product in a given context.

10. Application of Functional Analysis

Within laboratory M3M of the University of Technology of Belfort-Montbéliard, we carried out the application of the functional analysis (AFFUTEE, set-up by Pr. Garro O. and his collaborators), to the design of a drier for foodstuffs. That FA leads to the setting-up of Functional Specifications Textbook (CdCF).

With the help of this CdCF, we set up using the software CATIA, an overall drawing of the drier. This drier is presented in the figures showed below:

11. Findings

These drawings carried out due to the software of CAD (Computer-aided design), CATIA, make it possible to have an idea of the drier whose prototype remains to be constructed on the spot in Chad. We voluntarily omitted to define the energy source to make our system more flexible and adaptable to various sources of energy. The materials which we chose are available locally (sheets and wood primarily). Even the ventilator can be locally manufactured with the resources available on the spot. We propose in addition, to manufacture the first prototype in Chad by utilising local materials and technology. This is an example where inclusive innovation by taking local sources has taken precedence over technology transfer from outside. Much of the renewable energy will benefit from the knowledge, learning, innovation and capability building by researching, identifying, selecting, using, applying and maintaining local sources, both material, techniques and energy.

References

Adhikarya, R. (1994), 'Strategic Campaign Extension: With participatory-oriented method of agricultural extension', with box study of FAO' S experiments, FAO, Rome, Italy. Available at: http://www.FAO.org.

Akrich, M., Callon, M. and Latour, B. (1988), 'In what hold the success of the innovations? First episode: The art of the profit-sharing', in *Annals of the Mines*, June.

Akrich, M., Callon, M. and Latour, B. (1988), 'In what hold the success of the innovations? Second episode: Art to choose the spokesman', in *Annals of the Mines*, September.

Amin, A. (1974), *History of Islam* (in Arabic, 4 volumes), T.2 Editions: *Dar-el-Maarif*, Cairo, Egypt.

Delafollie G. (1996), *Analyse de la valeur*, Edition Hachette, Paris.

Desmorieux, H., Yacoub, I.H. and Garro, O. (2001), 'Characterization of Spirulina drying for design of dryers for developing countries', Proc of FirstNordic Drying Conference, organized by SINTEF Energy Research and Norwegian University of Science and Technology (NTNU), Editors Odilio Alves-Filho, Trygve Mr. Eikvik, Ingvald Strömmen, June 27-29.

Durand, C. (1994), *International technological co-operation, Technology transfers*, DE Boeck-wesmael S.A. Brussels.

Jeantet A. (1998), 'Les objets intermédiaires dans la conception in Eléments pour une sociologie des processus de conception', in *Sociologie du Travail*, 3: 291-316.

Rouach, D. and Klatzman, J. (1993), *Technology transfers: what do I know?* PUF.

Yacoub, I.H. (1998), 'Methodology of technology transfers towards the developing countries', Memory of DEA (MSC) of the Institut National Polytechnique of Grenoble, Higher National School of Industrial engineering, 57p.

Yacoub, I.H. (1999), 'Methodology of technology transfer', example of Spirulina, letter of research and development, CNAR, N°66, pp 1-2.

Yacoub I.H., Garro, O., Desmorieux, H. and Menguy, G. (2006), 'Technology transfer, a problem of innovation, which methodology for the technology transfers?' in *International Journal of Design and Innovation Research*, 3(3): 50-65.

Yacoub, I.H. (2010), *Methodology of North-South technology transfer: Development of Spirulina,* European University Press.

Acknowledgements

This work was partly supported by the University Agency of Francophonie (AUF). We thank the efforts of AUF for the development of scientific research in Africa.

Chapter 5

Reducing the Energy Bills of Higher Educational Institutions with Solar Energy:

The Case of the University of Ngaoundere in Cameroon

César Kapseu[1], *Noël Djongyang*[2], *Raoult T. Dabou*[3], *Maxwell T. Nsiéwé*[4], *Ibrahim Njukouyou*[5] *& Fabrice Debel Eyono*[6]

1 Department of Process Engineering, National School of Agro-Industrial Sciences, University of Ngaoundere, Cameroon, Email: kapseu@yahoo.fr
2 Department of Physics, Faculty of Science, University de Ngaoundere, Cameroon. Email:noeldjongyang@gmail.com
3 Department of Computer Science, Electronics and Automatics, National Advanced School of Agro-Industrial Sciences, University of Ngaoundere, PO Box 455 Ngaoundere, Cameroon. Email: dabour_fils3@yahoo.fr
4 Department of Energy and Process Engineering, National Advanced School of Agro-Industrial Sciences, University of Ngaoundere, PO Box 455 Ngaoundere, Cameroon. Email: maxwelltientcheunsiewe@yahoo.fr
5 Department of in Computer Science, Electronics and Automatics, National Advanced School of Agro-Industrial Sciences, University of Ngaoundere, PO Box 455 Ngaoundere, Cameroon. Email: nikhalil2000@yahoo.fr
6 Department of Agro-industrial Engineering, National Advanced School of Agro-Industrial Sciences, University of Ngaoundere, PO Box 455 Ngaoundere, Cameroon. Email: efabricedebel@yahoo.fr

1. Introduction

Technical progress and development are often connected with the quantity and the quality of energy used. Since the beginning of the industrial revolution three hundred years ago, electricity powered various field of activities, and the use of industrial energy has created an unprecedented improvement in average living standards and human well-being around the world (Bradford, 2006).

Humans rely heavily on energy utilisation for their every-day life and the energy demand is growing increasingly. Most of the countries in the world depend on fossil energies. But the conventional energy sources are limited and not available everywhere. The earth's reserves of coal, oil and gas are finite and likely to become resource-depleted in the course of this century (Archer and Hill, 2001). Moreover, these sources of energy generate a great quantity of polluting gases responsible of the global warming. The requirement for sustainable energy provision became increasingly important as the environmental effects of fossil fuel use become apparent (Smith, 2002; Schaeffer *et al.*, 2012). To address the problem of energy consumption while preserving a clean environment, it is necessary to look for new sources of energy (Schaeffer *et al.*, 2012). Solar energy appears to be a good alternative. It presents advantages of being sustainable (non-deplete-able), ubiquitous (found every-where across the world in contrast to fossil fuels and minerals), and essentially non-polluting compared to the fossil one. The promotion of increased research and development in the field of new energy service, including renewable energy, energy efficiency and advanced sustainable energy technologies, is necessity (Lam *et al.* 2004; Lacomelli, 2005; Singo, 2010). The last decades have witnessed the utilization of renewable energy sources in a large scale (Boubaker, 2012). Advantages offered by renewable energy resources have attracted much the attention of academics and professionals especially after Kyoto Protocol (Menegaki, 2008).

Energy saving is a high-priority worldwide. For this reason, energy-efficient measures are being increasingly implemented in all sectors. The residential sector is responsible for an important part of the energy consumption in the world (Pacheco *et al.*, 2012). In fact, the modern comfort living conditions are achieved at the cost of vast energy resources. The case of the University of Ngaoundere in Cameroon is an illustrating example. In fact, the electricity consumption of that higher educational institution is exorbitant (approximately 900 kWh/day) and increases year by year. The average daily energy cost is estimated to 486 US Dollars. This study aims to contribute in reducing the energy bills of the University of Ngaoundere through the use of solar energy. Both the technical and economic aspects of the project are presented.

2. Solar Energy

The earth receives each hour approximately 173 x 10^{12}kwh of energy from the sun. Over a year, this corresponds to 5 160 Q (1 Q = 1 055.06 x 10^{18} J) (Neville, 1995). However, not all this energy reaches the surface of the earth. A portion amounting to some 1 570 Q is reflected by clouds, the oceans and the land. The energy from the sun can either be used directly (solar thermal) or transformed into electricity (solar photovoltaic) through solar panels.

The key elements of photovoltaic (PV) conversion are solar cells. The photons reaching a solar cell move the electrons in the material, and thus generate a continuous flow of electrons (electric current). However, the efficiency of a PVcell depends on some environmental parameters like solar irradiation, and temperature distribution at the surface of the cell. Thus, each current-voltage (I-V) curve corresponds to a specific operating condition. Association of PV cells permits to get a desired voltage and/or current (Bernard, 2004). The basic installation of a PV system is presented in Figure 1.

Energy supply from the sun significantly contributes to reduce the production of electricity from nuclear or fossil sources of energy which are not environmentally clean. The first type can generate radioactive elements hazardous to human beings (e.g. Chernobyl in 1986 and Fukushima in 2011) while the second one emits important amounts of polluting gases responsible of the global warming.

Figure 1: Basic Installation of PV Systems

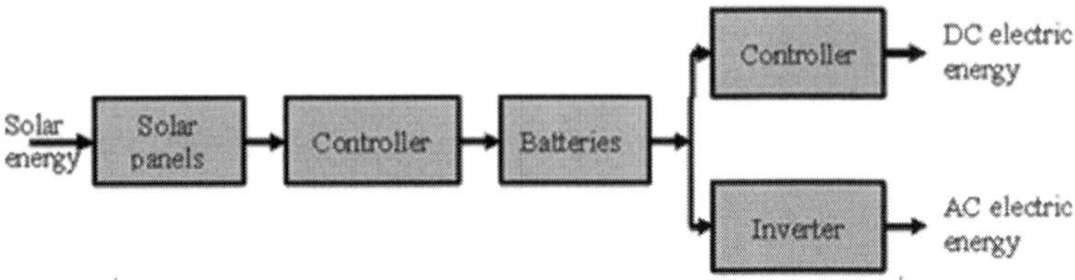

3. Energy use at the University of Ngaoundere

The study Area

The University of Ngaoundere (7° 19′ N, 13° 35′ E) is one of the eight State Universities in Cameroon. It is in the region of Adamawa in the northern part of the country, and made up of:

- Four faculties: the faculty of sciences; the faculty of laws and political sciences; the faculty of arts, letters, and social sciences; and the faculty of economic sciences and management;
- Three engineering schools: The University Institute of Technologies, the National School of Agro-Industrial Sciences; and the School of Geology and Mining;
- A school of medicine and veterinary sciences.

Its geographical position is a major asset for an effective exploitation of solar energy. Table 1 presents the mean solar irradiation on the site, evaluated through PVGIS (http://sunbird.jrc.it/pvgis). The mean value obtained (5.9 kWh/

m^2/day) is not far from the 5.78 kWh/m^2/day found in some literatures (Nkué and Njomo, 2009; Ayangma *et al.*, 2008).

Table 1: Monthly Mean Solar Irradiation over a Flat-Horizontal Surface in Ngaoundere

Months	Jan	Feb	Mar	Apr	May	Jun	Jul	Aug	Sep	Oct	**Nov**	**Dec**
Irradiation (kWh/m^2/day)	6.79	7.03	6.81	6.16	5.63	5.28	4.94	4.86	5.03	5.48	6.38	6.48
Mean yearly value in kWh/m^2/day								5.9				

Source: http://sunbird.jrc.it/pvgis

The Energy Consumption of the Institution

Table 2 gives an overview of the current status of energy consumption at the University of Ngaoundere. It presents the monthly electricity consumption (in kWh) for two years (2008 and 2011).

Table 2: Monthly Electricity Consumption at the University of Ngaoundere in 2008 and 2011

	Electricity Consumption (kWh)	
Months	**2008**	**2011**
January	/	19056
February	30955	17964
March	27414	15229
April	28949	15862
May	12836	12653
June	45364	16913
July	51949	14571
August	34101	10715
September	31802	10566
October	30673	10738
November	12327	15021
December	13201	17861

It could be seen that the energy consumption of the institution is exorbitant and increases year by year. This could be easily checked in terms of money (see Table 3). However, most of the energy is consumed by the University Institute of Technologies and the National School of Agro-Industrial Sciences. In fact, these two engineering schools produce some foods and drinks such as bread, cakes, yoghurt, ice-cream, etc. for both practical and commercial purposes. In addition, they have a small brewery unit.

Table 3: Electricity Bills of the University of Ngaoundere in 2008 and 2011

	Electricity bills (in US Dollar)	
Months	**2008**	**2011**
January	/	17127
February	30159	16374
March	13330	14376
April	13286	16661
May	13432	12840
June	15120	15434
July	13781	14392
August	11262	12905
September	10636	12009
October	10316	12500
November	12413	14615
December	12806	15594

From the above, it could be checked that the average daily expenditure is about 486 US dollars while the average daily energy consumption is 969 kWh. The question one can ask is: "if the electricity need has to be supplied by solar panels, what is the size the suitable PV system?" The following section tries to answer the question.

4. Sizing the PV System

Material

In this study, polycrystalline PV panels (type BP 3125U, 125 W–24 V) manufactured by Helios Technology© are used because of their cost, efficiency, and availability. The energy storage system is made-up of an association of ACD Industrial Batteries accumulators (12 V, 166 Ah). The system is assumed to provide 1868.0495 Ah, with autonomy of 2 days.

Calculations

Energy produced by the photovoltaic system

The energy produced by the PV system is obtained through the formula:

$$E_{pv} = E_{load}/k \tag{1}$$

where E_{pv} (in Wh) is the energy produced by the PV system, E_{load} (in Wh/day) is the daily load, and $k \approx 0{,}55–0{,}65$ is a corrective factor including the overall outputs of the system (a mean value of $k = 0.6$ was taken in this study).

Calculations give $E_{pv} = 1489.8415$ Kwh. The size of the suitable generator can be deduced from that result and its value is 395.1834 kilowatts.

Number of PV panels

The number of the PV panels is obtained through:

$$N = E_{load}/\left(0.8 \times E_{pv}\right) \tag{2}$$

Where N is the number of PV panels, E_{load} (in Wh/day) is the daily load, and E_{pv} is the energy produced by the PV array. Calculations give N = 3 162 panels

Battery Capacity

Thebattery capacity is:

$$C = (E_{load} \times DA) / (DD \times U \times \ _{battery}) \quad (3)$$

C (in Ah) is the capacity of the battery; DD is the depth of discharge of the accumulator (in general 50 % ≤ DD ≤ 80 %; DD is assumed to be 80 % in this study), DA is the autonomy duration; U (in V) the nominal voltage of the accumulator, and $\eta_{battery}$ is the efficiency of the battery (assumed to be 85 %).

Calculations give C =44833.19Ah, i.e.271branches of accumulators.

Maximum current between the batteries and the inverter

Calculations give the maximum current between the batteries and the inverter as:

$$I_{max} = P_{max,inverter} / U_{battery} = 95.8 \text{ A} \quad (4)$$

where I_{max} (in A) is the maximum current between the batteries and the inverter, $P_{max,inverter}$ (in W) is the maximum power consumed in the inverter, and $U_{battery}$ (in V) is the voltage of the battery.

Cables Section

The section of the cables is obtained by:

$$S l = R \, / \quad (5)$$

where ρ is the resistivity (in Ω.m) of the cables, l the length of the cables (in m), and R (in Ω) the electrical resistance.

The section of the cables can also be chosen in tabulated tables. TH (122 mm^2) cables are used between the junction box and the inverter, while TH (35 mm^2) cables are used between the battery and inverter.

Technical Maintenance

Figure 2 provides the FAST (*Functional Analysis System Technique*) diagram below which shows the technical operations to maintain the PV system.

Figure 2: FAST Diagram

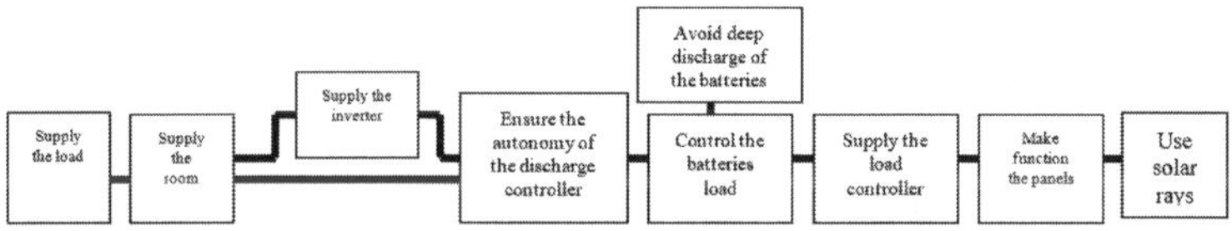

Table 4: Estimated Cost of the System

Designation	Quantity	Unit price (US Dollar)	Total (US Dollar)
Solar panel	3162	137	43210627
Controller 24 V	56	1367	7652736
Battery	463	64	2920234
electrical protection	32	631	2018304
Inverter-Charger	172	1701	29249632
Connecting cables and electrical accessories		/	62219952
Others			6070000
Total			**153301485**

Payback Time (PBT) of the Investment

The maximum daily consumption of about 969 kWh per day corresponds to an expenditure of 486 US Dollar per day with the national operator of electricity (AES-Sonel) as energy supplier i.e. 174833 US dollars for a whole year. The subscription fees amount to 20 US Dollar/year. The total amount is therefore TA_{AES} = 174853 US Dollar/year. The payback time is therefore given by:

$$PBT = TA_{PV} / TA_{AES} \qquad (6)$$

where PBT (in years) is the payback time of the investment, TA_{PV} (in US dollars/year) is the total amount of the PV system and TA_{AES} (in US dollar/year) is the total amount of the system with AES-Sonel as energy supplier. Calculations gives PBT=8.76 years.

5. Findings

The findings from this feasibility study of energy supply with solar energy at the University of Ngaoundere in Cameroon show that, with an irradiation of 5.9 kWh/(m^2.day) and an estimated need of 969 kWh/day, 3162 solar panels (125 W, 24 V) would be necessary, to cover the energy needs of the institution. The overall investment would be 153301485 US dollars and the payback time of the investment is approximately nine years. A total gain of 3671893 US dollars is performed, if the system is assumed to function for thirty years.

References

Archer, M.D. and Hill, R. (2001), 'Clean electricity from Photovoltaics', in *Series on Photoconversion of Solar Energy* - Vol. 1, Imperial College Press.

Ayangma, F., Nkeng, G.E., Bonoma, D.B. and Nganhou, J. (2008), 'Evaluation du potentiel solaire au Cameroun: cas du Nord Cameroun', in *African Journal of Science and Technology (AJST) Science and Engineering series,* 9(2): 32-40.

Bernard, J. (2004), 'Energie solaire: calculs et optimisation', Ellipses.

Boubaker, K. (2012), 'Renewable energy in upper North Africa: Present versus 2025-horizon perspectives optimization using a Data Envelopment Analysis (DEA) framework', in *Renewable Energy,* 43: 364-369.

Bradford, T. (2006), *Solar Revolution: The Economic Transformation of the Global Energy Industry,* The MIT Press. Available at: http://sunbird.jrc.it/pvgis(Accessed: 07 August 2012).

Lacomelli, A. (2005), 'Renewable Energies for Central Asia Countries: Economic, Environmental and Social Impacts', in *Series IV:Earth and Environmental Sciences* – vol. 59, NATO Science Series, Springer.

Lam, J.C., Chan, R.Y.C., Tsang, C.L. and Li, D.H.W. (2004), 'Electricity use characteristics of purpose-built office buildings in subtropical climates', in *Energy Conversion and Management,* 45: 829–844.

Menegaki, A. (2008), 'Valuation for renewable energy: A comparative review', in *Renewable and Sustainable Energy Reviews,* 12: 2422 - 2437.

Neville, R.C. (1995), *Solar Energy Conversion: The Solar Cell,* Elsevier Science Store. Available at: http://store.elsevier.com/Solar-Energy-Conversion/R_C_-Neville/isbn-9780444898180/

Nkué, V. and Njomo, D. (2009), 'Analyse du système énergétique camerounais dans une perspective de développement soutenable', in *Revue de l'Energie,* vol. 588, Mars-Avril.

Pacheco, R., Ordóñez, J. and Martínez, G. (2012), 'Energy efficient design of building: A review', in *Renewable and Sustainable Energy Reviews*, 16: 3559-3573.

Schaeffer, R., Szklo, A.S., Pereira de Lucena, A.F., Borba, B.S.M.C., Nogueira, L.P.P., Fleming, F.P., Troccoli, A., Harrison, M. and Boulahya, M.S. (2012), 'Energy sector vulnerability to climate change: A review', in *Energy*, 38: 1-12.

Singo, A.T. (2010), 'Système d'alimentation photovoltaïque avec stockage hybride pour l'habitat énergétiquement autonome, *PhD Thesis*, University Henri Poincaré, Nancy.

Smith, N.A.V. (2002), 'Decision Support for New and Renewable Energy Systems Deployment', *PhD Thesis*, University of Strathclyde, UK.

Chapter 6

Design of an Electronic Power and Charge Controller for Hybrid Energy Systems (EPOCHES)

André R. Tchamda[1], *Robert Tchitnga*[2], *Christian T. Tchapga*[3] *& Martin Kom*[4]

1. Introduction

The global awareness and concerns of the pollution caused by the use of polluting energy sources spurs the general tendency to promote the development and

1 Research Group on Experimental and Applied Physics for Sustainable Development, EAPhySuD, University of Dschang, Cameroon, Email: rodriguetchamda@yahoo.fr

2 Research Group on Experimental and Applied Physics for Sustainable Development, EAPhySuD, University of Dschang, Cameroon. Laboratoire des Energies Renouvelables, LER, Université de Dschang, B.P. 313 Dschang, Cameroon. Email: tchitnga@yahoo.fr

3 Research Group on Experimental and Applied Physics for Sustainable Development, EAPhySuD, University of Dschang, Cameroon; College of Technology, Department of Electrical and Electronic Engineering, University of Buea, Cameroon. Email: ctchito@yahoo.fr

4 Groupe de Recherche sur les Technologies Médicales Adaptées aux Tropiques, Laboratoire d'Electronique et de Traitement du Signal, Département des Génies Electrique et Télécommunications, ENSP, Université de Yaoundé I, Cameroun. Email: kommart@yahoo.com

the use of clean energy. This development will be growing exponentially, for reasons including the global warming resulting from the use of polluting energy sources like fossil and fissile, and the long-term risk of a shortage of these old energy sources (Vitousek, 1994). The production of electricity from renewable energy sources such as the sun, the wind or the biomass is therefore gaining importance. In the coming decades, renewable energy will contribute significantly to global electricity production and sustainable development (Fiedermann *et al.*, 2012). Moreover, the accident at the nuclear base of Fukushima, Japan in 2011 came to sound the alarm about risks of nuclear pollution in addition to the threats of climate warming from the global greenhouse effect (Boulton, 2011).

The storage of electrical energy produced by fluctuating and irregular renewable energy sources poses technical requirements of several types. Among others, we can name the design of storage elements like batteries, and the design of electric storage modules for the management and use of this energy.

Hybrid systems of energy production in their most general view are that which combine and exploit several available sources of different nature at the same time. To make use of combined energy sources (e.g. wind-solar-diesel) to provide electricity needed to power technical devices, electronic modules for the management of the so-called hybrid energy systems have been developed (Saheb-Koussa and Belhamel, 2007). Most of these modules are used to control and manage batteries charging from all available renewable energy sources. Many charge controllers existing on the market perform their controlling function based on the voltage or on the current delivered by the renewable energy sources to the batteries (Saheb-Koussa and Belhamel, 2007; Tangka *et al.*, 2010). In the case of solar energy, considerable works have been done in which the tracking of the maximal power delivered by the panels of photovoltaic system according to the irradiation was the central point for the optimization of charging batteries. These include works such as Toshihiko *et al.* (2002), Veerachary *et al.* (2003), Bahgat *et al.* (2005), Hussein *et al.* (2005), Salas *et al.* (2006), Ouariachi *et al.* (2009) and Yaden *et al.* (2011). However, this maximal power of the panel refers to the power upstream of the batteries and not the power requested by the loads connected to these batteries.

The long life of batteries depends among others on the characteristics of the load connected to it. Yet, it is known that if the total power requested by different loads connected to the batteries is greater than the power upstream of it, that is to say, power brought by the renewable energy sources to the batteries, there will be a rapid and deep damageable discharge of these batteries. The present work aims at proposing a solution to prevent batteries from deep discharge nuisance by using the proposed EPOCHES which plays simultaneously the role of power controller and charge controller.

2. Measurement and Control of Signals (Voltage and Power)

Principle of Reading the Voltage of the Battery by the Microcontroller

The design controller is made for loads of average power consumption of alternative current by converting the direct current using an inverter. Given inconstancy of renewable energy sources such as the wind and the sun, this controller will be equipped with a diesel engine input too, making it hybrid. The combination of many renewable energy sources optimizes at maximum the power generation systems. The optimization of wind energy, photovoltaic and diesel by electrochemical storage has allowed the establishment of charge controllers, energy distribution and batteries protection (Tangka *et al.*, 2010; Toshihiko *et al.*, 2002; Veerachary *et al.*, 2003; Bahgat *et al.*, 2005). These controllers do not incorporate the process of controlling the power needed by the loads connected to the batteries (Ouariachi *et al.*, 2009). For this reason, we propose a controller (EPOCHES) which not only integrates all the functions of other controllers (charging control and discharging batteries, voltage regulation etc.), but also the power controller function of the energy provided to the load by the batteries. It needs to be noted that the EPOCHES can only control a certain power range.

The maximal value of the power P_S that the batteries can deliver to the loads will be defined at the starting of the module, using a keyboard data entry. This value will be stored in the EEPROM of the microcontroller. Thus, the microcontroller can calculate the electrical efficiency η by solving the equation: $\eta = \frac{P_a}{P_S}$, where P_a is the power absorbed by the loads.

Now, knowing the calculated performance will help to see whether there is good demand for power. Thus, the threshold performance is set at 0.95 that is to say that the EPOCHES can only let 95% of the power of the batteries to pass to the loads. The remaining 5% will be used to maintain $P_s > P_c$. If the performance $\eta > 0.95$, a warning alarm and a light will be triggered by the microcontroller and the power source will be disconnected (we used the electrical network of the house (loads). With the aid of a LCD screen from a recycled telephone screen of type NOKIA 3310), the module will display the pattern of power failure.

The microcontroller PIC16F877A used in the device has analog inputs and analog-digital converters (Microship, 2012). The reference voltage of the PIC that we selected for analog to digital conversion (ADC) is $1.2V$. But we want this PIC to measure voltage across the battery, which is set to 14.4V at maximum. This voltage being greater than the reference value, we cannot connect it to the PIC, for fear of damaging it or, in the best, causing its malfunction that can result e.g. in an incorrect reading of the different variations of the voltage across the battery. A voltage divider bridge is necessary to adapt the battery voltage to the reference voltage of the PIC as defined in the formulas below (see Figure 1).

For a voltage $U = 14.4V$ across the batteries, if $R_1 = 2k\Omega$ and $R_3 = 22k\Omega$, we obtain the convenient voltage $V = 1.2V$ for the PIC, which is equal to the reference voltage. The PIC may well observe variations in the voltage at the terminals of the battery. Thus, after reading, the PIC will divide the reading by the coefficient k given above to obtain the actual value of the voltage.

Figure 1: Voltage Divider and Formulas

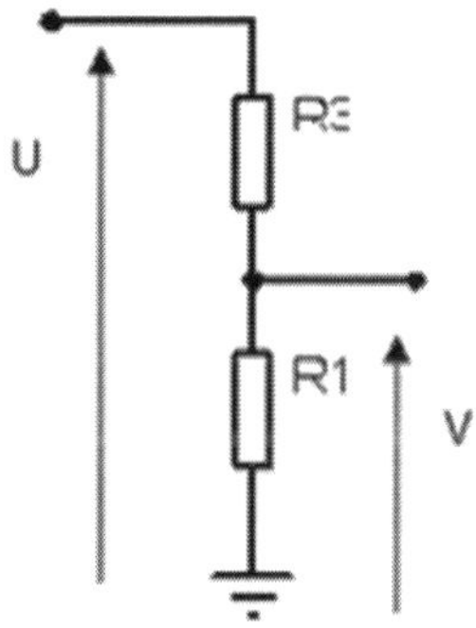

$$V = \frac{R_1}{R_1 + R_3} U \qquad k = \frac{R_1}{R_1 + R_3}$$

The Measuring Principle of Power by the Microcontroller

The power is given by $P = U \cdot I$ this is understandable since the reading device of the power is located between the batteries and the inverter; that is to say on the side where the signals describing the energy are still continuous (DC). To determine the required power, we simply have to know the voltage across the battery and the current passing through it. A $12V$ battery consists of 6 cells or compartments. Generally, we consider that a lead-acid battery is discharged when the voltage per cell is less than $1.8V$. A $12V$ battery is said discharged when the voltage at its terminals is less than or equal to $10.8V$. The minimum value of 10.8V across the batteries being defined, it remains to determine the value of the current.

To read the value of current in the circuit, we use the shunt (Taiwan trade, 2011). The current sensor used is a shunt resistor with characteristics $(100A/100mV)$ (Taiwan trade, 2011). For a current of $100A$ flowing through the shunt resistor, a potential difference (pd) of $100mV$ across the resistor, from where the ratio $R = 0.001\Omega$. This ratio is very typical for a current measurement by microcontrollers. As the microcontroller cannot measure the voltage, a measurement of current is required, followed by a current to voltage conversion via the application of the Ohm's law. The microcontroller can calculate the current in the circuit and multiply it by pre-set voltage to have the power absorbed. An example of diagrams illustrating the current reading is given by Figure 2.

Figure 2: Block Diagrams Illustrating the Technique of Current Reading

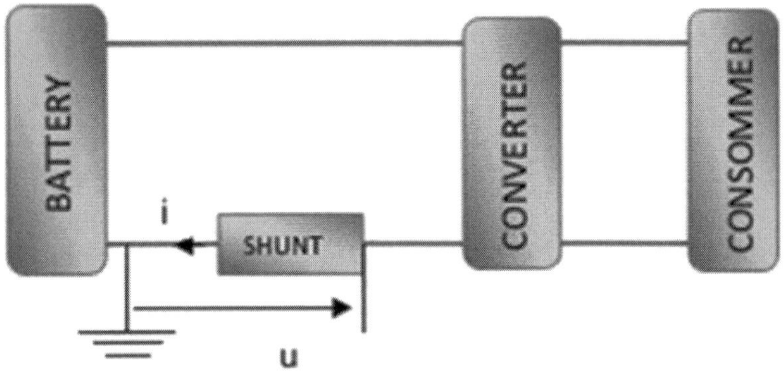

Taking into consideration the characteristics of the shunt resistor $(100A/100mV$ and the operating range specified by the manufacturer which is 66% of the maximal current of $100A$ for the chosen resistor, the current through the shunt will not exceed $66A$. The range of current flowing through the shunt resistor must be respected. It imposes a power range that the device can read and control without deteriorating.

The maximum voltage drop $(100mV)$ across the shunt resistor is negligible for a current limit of $100A$. This is less than the voltage reference of the PIC; the microcontroller can therefore observe and follow the variations of current in the power circuit.

Control of the Charging Voltage

A lead battery contains elements (cells) of approximately 2V each, that loads up to 2.4V. A12V battery has 6 of such cells. It then charges to a maximum voltage of 14.4V. In our control scheme below (see Figure 3), the integrated circuit U47812 helps to regulate the voltage at 12V. Now we need a voltage up to $14.4V$ to charge the 12V battery. For this, four $1N4001$ diodes of threshold voltage $0.6V$ mounted in series are used, for a maximum threshold voltage of $2.4V$. These diodes associated with the regulator, provide an output voltage of $12V + (0.6V \text{ x } 4) = 14.4V$, which is enough to charge the battery.

The transistor Q1 conveys a strong 'current up to 8A in the rest of the circuit without deteriorating the electronic component.

Figure 3: Schematics of the Electrical Voltage Regulator

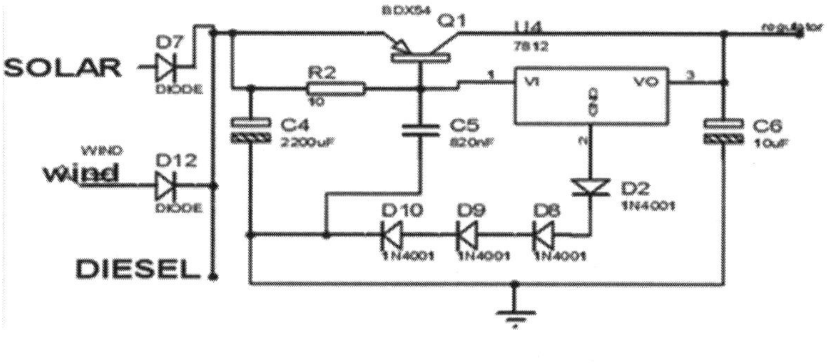

3. Determination of the Power Range

To determine the power, we need two parameters: the voltage and the current. A $12V$ lead battery is said to be discharged when the voltage across it is less than or equal to $10.8V$. This value will be recorded in the memory of the PIC as the threshold value. Since the battery drains during use, it will not be wise to set the battery voltage in the memory of the PIC up to $12V$. The voltage to be set in memory is $10.8V$. It remains to determine the current value. For the security of the shunt resistor we will use 66% of the current (Taiwantrade, 2011). The limit value of current to consider is therefore $100A \text{ x } 66\% = 66A$.

$$P = U \cdot I \qquad ====> \qquad P = 10.8V \text{ x } 66A = 712W$$

Thus, the variation range of the power goes from $0W$ to $712W$. We chose to control the power in continuous operation. It is very clear and simple to use microcontroller to read a positive voltage. Factors such as $\cos(\varphi)$ will not be taken into consideration in the framework of this paper, since it is not constant for all the users (loads), particularly when in alternative regimes inductive and capacitive loads are present.

Controller Architecture and Operation of the EPOCHES

Figure 4 displays the block diagram of the EPOCHES. Important blocs can be seen: On the left side, the hybrid inputs are presented. The PIC piloting the whole system thrones in the center of the map.

Initially, each wind turbine and solar panels provide power generation which will be used to charge batteries. They are all connected to the device by solar and wind inputs. A diesel entry is foreseen in case of no wind and sun. All these inputs are connected to the voltage regulator that will keep the charge voltage to 14.4V if the voltages produced by the wind, photovoltaic or diesel are higher than the charging voltage. The diesel generator is connected to a voltage rectifier that converts the AC voltage into DC voltage for charging batteries. The LCD screen is the interface between the operator and the module. It can display physical values such as: the power consumed by the loads (habitat); the voltage at the battery electrodes and so on. The keyboard allows the power set necessary for batteries to charge. The inverter converts the direct voltage into alternating voltage for supplying the operating AC loads. Switches RL1 and RL2 are used to connect the loads either at the output of the inverter when the batteries are charged, or to the provided diesel generator if the batteries are discharged. The switch RL3 can disconnect the battery from the generators when the microcontroller wants to read the voltage across the battery. This is to avoid reading but the voltage across the generators (photovoltaic, wind, diesel), as the battery is normally connected to the generators during the charging process. It makes sense to disconnect generators. The part consisting of integrated circuits (7012 and 7805) is the power supply which provides energy to the rest of the system. In the middle of the figure, we have a microcontroller PIC16F877A which is the control center of the whole module. It is in this circuit that the controlling program for the voltage across the battery, for current and power supplied to the load by the batteries is embedded.

Figure 4: Block diagram of the EPOCHES

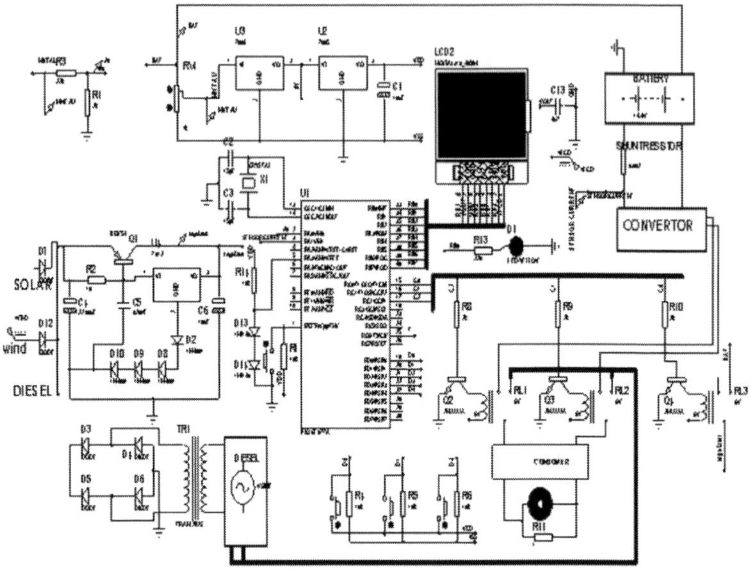

Legend

- Power supply: it distributes the amount electricity needed in the various blocks of the controller.

- Detector of voltage level: it is the interface between the electronic module and battery management. Its role is to measure the voltage level of battery terminals and to convert it into a voltage understandable by the microcontroller. This voltage is read by the PIC to inquire about the level of battery voltage.

- Solar and wind power input: it allows the energy available at the terminals of PV or windwheel and transmit it to the control system.

- Regulator: helps to regulate to a maximum voltage level $(15V)$ the voltage coming from the rectifier and the voltage at the output of solar and wind system entry.

- Rectifier: converts the AC voltage coming from the emergency generator in a continuous voltage and sent it to the interface relay that will be used for battery charging or for consumption as interpreted by the microcontroller.

- Relay interface: is the channel through which orders are translated into physical actions in terms of equipment that are actually ordered.

4. Findings

Another approach has been introduced regarding charge controllers, which is usually not taken into account, namely the power control of the energy delivered by batteries from renewable power stations to the loads instead always of only the voltage control at the charge and discharge of these batteries. The advantage of this approach is to prevent voltage drops and the damaging deep discharge of batteries that are sometimes faced in such systems. A short description of the module has been given and the global block diagram presented. The findings show that 'the EPOCHES integrates the functionalities of the previous controllers and complete it by this new option.

References

Bahgat, A.B.G., Helwa, N.H., Ahmad, G.E.and El Shenawy, E.T. (2005), ' Maximum Power Point Tracking Controller for PV Systems Using Neural Networks', in *Renewable Energy*, 30(8): 1257-1268.

Boulton, F. (2011), 'The Fukushima Nuclear Disaster: What may we learn from Fukushima?' *Medicine & Global Survival*, Special Edition, 57 -60.

El Ouariachi, M., Mrabti, T., Tidahf, B., Kassmi, Ka.and Kassmi, K. (2009), 'Regulation of the Electric Power Provided by the Panels of the Photovoltaic System', in *International Journal of Physical Sciences*, 4(5): 294 -309.

Fiedermann, S., Halilovic, J. and Bogacz, T. (2012), '*Solar potential of the Sahara Desert with an introduction to solar up-*

draft power plants', Available at: http://eeeic.org/proc/papers/129.pdf.

Hussein, K.H., Muta I., Hoshino, T. and Osakada, M. (2005), 'Maximum Photovoltaic Power Tracking: An Algorithm for Rapidly Changing Atmospheric Conditions', in *IEE Proceedings Transmission and Distribution*, 142(1): 59-64.

Microship (2012), *Analog/Digital ConvertorDatasheet of PIC16F877*, Available at: http://ww1.microchip.com/downloads/en/devicedoc/30292c.pdf.

Saheb-Koussa, D.and Belhamel, M. (2007), *Système hybride (éolien – photovoltaïque – diesel) à fourniture d'énergie électrique totalement autonome*, Revue des Energies Renouvelables CER'07 Oujda 67-70.

Salas, V., Olıas, E., Barrado, A. and Lazaro, A. (2006), 'Review of the Maximum Power Point Tracking Algorithms for Stand-Alone Photovoltaic Systems', in *Solar Energy Materials and Solar Cells*, 90(11): 1555-1578.

Taiwantrade (2011), *Types of shunt resistors*, Available at: http://www.taiwantrade.com.tw/MGSswitches/productsdetail/en_US/162001/DC_Shunt(100A_100MV%29).

Tangka, J.K., Tchakoua, P., Fotsin, H. and Fomethe, A. (2010), 'Conception et réalisation d'un module électronique de contrôle de charge et de gestion optimale de l'énergie pour systèmes énergétiques hybrides éolien-diesel, photovoltaïque-diesel et éolien-photovoltaïque-diesel(MEC-CGOPSEH)', in *Revue des Energies Renouvelables*, 13(4): 591-602.

Toshihiko, N., Togashi, S. and Roy, N. (2002), 'Short-current Pulse-based Maximum-Power-Point *Tracking Method f*or Multiple Photovoltaic-and-Converter Module System', in *IEEE Transactions on Industrial Electronics*, 49(1): 217-223.

Veerachary, M., Senjyu, T. and Uezato, K. (2003), 'Neural-Network-Based Maximum-Power-Point Tracking of Coupled-Inductor-Interleaved-Boost-Converter-Supplied PV

System Using Fuzzy Controller', in *IEEE Transactions on Industrial Electronics*, 50(4): 749-758.

Vitousek, P.M. (1994), 'Beyond Global Warming: Ecology and Global Change', in *Ecology*, 75(7): 1861-1876.

Yaden, M.F., El Ouariachi, M., Mrabti, T., Kassmi, Ka., Tidhaf, B., Chadli, E. and Kassmi, K. (2011), 'Conception et réalisation d'un système photovoltaïque muni d'une commande MPPT numérique', in *Revue des Energies Renouvelables*, 14(1): 171-186.

Chapter 7

The Status of Solar Photovoltaic and Solar Thermal Applications in Malawi

Chifundo M. Tenthani[1], *Esther Phiri*[2] *& Thomas Senganimalunje*[3]

1. Introduction

Malawi is a landlocked country that lies at 13°30′ S, 34°00′ N and is bordered by Tanzania in the North East, Zambia in the North West and Mozambique in the Southern part. The population of the country is 13 677 160 with an annual growth rate of about 3% (MDHS, 2010). Of this population about 85% live in the rural area where only 1% of that population has access to electricity. The country is one of the Least Developed Countries (LDC) and is ranked 168 out of 174 countries on the Human Development Index. The economy of Malawi is predominantly agriculture based with tobacco as the main foreign currency earner followed

1 Physics and Biochemical Sciences Department, The Malawi Polytechnic, Blantyre, Malawi. Email : ctenthani@poly.ac.mw

2 Electrical Engineering Department, The Malawi Polytechnic, Blantyre, Malawi. Email: ephiri@poly.ac.mw

3 Physics Department, The University of Malawi, Zomba, Malawi. Email: tmalunje@yahoo.com

by tea and cotton. With a land area of about 94, 000 square kilometres, it is one of the most densely populated countries in Africa with the population density hovering around 158.8 people per square kilometer.

Provision and access to modern forms of energy is still lacking in most countries including Malawi (UNDP, 2007). Malawi's energy balance is heavily dominated by biomass resulting in excessive use of fuel wood and charcoal as a source of energy both domestically and in industries (Zingano, 2005). This has resulted in massive deforestation even in those areas that act as catchment areas to rivers that are used in hydro generation. It also results in a lot of unproductive time for men and especially women because of having to walk long distances in search of firewood. Indoor air pollution is a common occurrence and it is known to cause adverse health effects (Wambua, 2011).

This study presents the current status of solar energy applications in Malawi. It will mainly look at the governance of the renewable energy sector, the general energy profile for Malawi and finally look at some solar energy case studies. The data in this paper was collected through consultations with line ministries in the energy sector, questionnaires, face to face interviews and through site visits to the places where the solar applications were installed.

2. Energy Sector Governance

The Department of Energy (DOE) in the Ministry of Energy and Mines is responsible for all the issues pertaining to energy in Malawi. To facilitate energy sector governance, the government established a sector-wide regulator known as the Malawi Energy Regulatory Authority (MERA) in 2008. The establishment of MERA came about following the merger of the Petroleum Control Commission (PCC) and the National Energy Council (NECO), which were overseeing the regulation of petroleum and electricity industries, respectively. MERA regulates the entire energy sector by developing and enforcing performance and safety standards for energy exploitation, production, transportation, and distribution.

The Malawi National Energy Policy (2003) was formed through a consultative and interactive process. It was developed from previous policy guidelines and strategies which were spread across several documents and was checked against the country's other policies namely: Malawi Vision 2020, Malawi Poverty Reduction Strategy Paper (MPRSP), and the UN MDGs (UNDP, 2007). The overall policy of the energy sector is to ensure adequate and sustained supplies of energy for continued economic growth and development. The energy policy recognises the dependence of Malawi on biomass and on imported energy sources, especially petroleum products. It also recognises the importance of developing and utilising indigenous energy resources, especially renewable energy such as biomass, solar, wind, mini/micro hydro, and geothermal energy.

In September 1999 the Government launched a National Sustainable and Renewable Energy Programme (NSREP) to promote a wide utilisation of RETs. The approach was to address the identified barriers: institutional, financial, technical, and standards. Since the inception of NSREP there have been several RETs projects ranging from policy initiatives to practical installations. The policy part focused on addressing the identified barriers while the practical part focused on demonstration installations and pilot projects. This programme saw the birth of the Barrier Removal to Renewable Energy in Malawi (BARREM) project which was implemented during the years 2000 to 2008. BARREM concentrated on solar energy technologies. Other RETs were implemented through other partners and government developmental projects.

3. Energy Provision

Malawi is endowed with several forms of energy resources, which include biomass, coal, a number of perennial rivers for hydropower generation, high solar radiation for photo voltaic and solar thermal applications, a reasonable wind resource for water pumping, hot springs for geothermal power generation, a good number of domesticated animals for biogas applications and uranium deposits, even though this is for export only.

The energy balance of Malawi has been dominated by biomass in terms of production and by the household sector in terms of consumption (GoM, 2003). Malawi's energy balance has been heavily biased towards wood fuel (93%) with petroleum, electricity, coal and other renewables accounting for only 3.5%, 2.3%, 1 % and 0.2%, respectively (see Figures 1 and 2). About 90% of the population depends on wood fuel for their thermal services, while paraffin (kerosene) and candles are used for lighting. There has been a slight change in the statistics over the years with biomass accounting for 88.5% of the total energy demand. The remaining part is supplied from hydrocarbons, 6.4%, electricity, 2.8%, and coal 2.4%. Data for renewable and alternative sources of energy was not available.

Figure 1: Malawi Energy by Sources

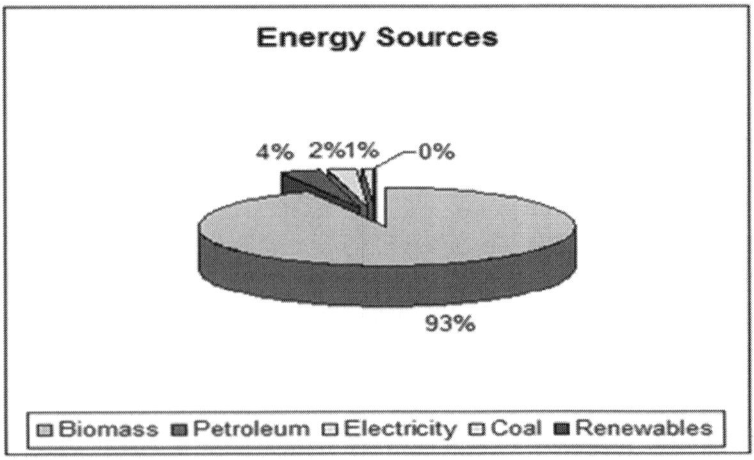

Source: GoM (2003)

Figure 2: Malawi's Energy by Consumption

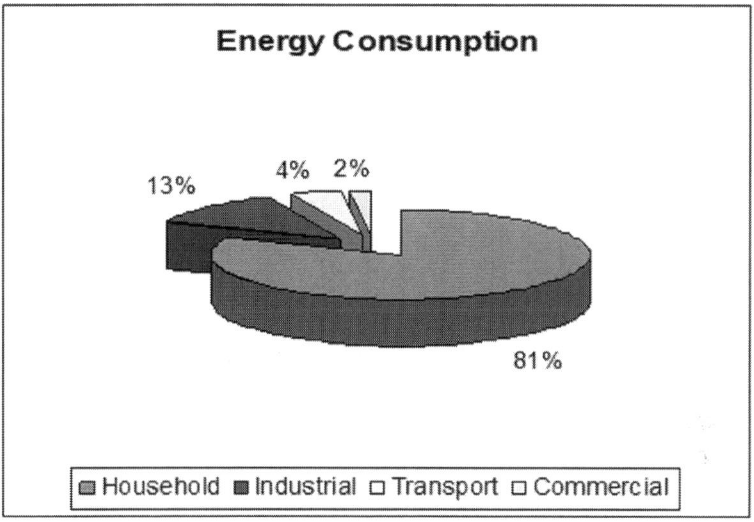

Source: GoM (2003)

Electricity

National electricity supply dates back to the 1950s with a coal-powered plant at what is now called ESCOM Powerhouse in Blantyre and a mini-hydro at Mulunguzi River in Zomba, the then capital of Malawi. The Malawi electricity supply industry is run by the publicly owned power utility; the Electricity Supply Corporation of Malawi (ESCOM). ESCOM is the sole provider of electricity and manages the generation, transmission and distribution of the power throughout the country

About 95% of Malawi's electricity needs are provided by hydropower using the centrally located Shire River cascaded hydro schemes, which provides 98% of the power and Wovwe mini hydro plant in the Northern Region for the remaining 2% (GoM, 2010). The total installed capacity of 302MW of which about 20% is lost during transmission due to resistive losses is not enough to meet the current demand. Even though only 7.6% of the Malawi population has access to the grid, load shedding is the older of the day. This figure accounts for 20%

of the urban population and only 1% of the rural population (Tenthani *et al.*, 2011).

Petroleum Products

Malawi has no oil production, known reserves, or refineries, making the country completely dependent on imports of refined oil products. There are no oil pipelines into Malawi and thus all of the country's petroleum products are imported and transported within the country by road tankers, and the national storage capacity is for only 60 days (GoM, 2008).

Petroleum, which includes oil and gas, represented 3.5% of the total energy primary consumption in 2003 of which 97% was imported and the remaining 3% was contributed by locally-produced ethanol, which is blended with petrol on a maximum 20:80 ratio of ethanol to petrol. The major petroleum consumer in the country are; the transport sector (90%), industries (2.8%), agriculture (1.9%) with about 5% being used for lighting in the form of paraffin (GoM, 2010)

Coal

Total proven coal reserves in Malawi are 1 billion metric tonnes, of which 22 million are proven reserves of bituminous type coal. The coal has a calorific value ranging from 28 to 30 MJ/Kg. Coal mining started in 1985 at the Mchenga coal mine in Livingstonia which is the main coal mine and has higher quality coal. It is also mined at Kaziwiziwi coal mine and Mean Jalawe mine opened in 2005. The production of coal in Malawi is increasing from 52, 000 tons in 2005 to 59,000 tons in 2009 (MSOER, 2010). In 2003, coal represented only 1% of the total primary energy consumption because coal resources were never seriously exploited. Malawi has been exporting coal to the southern region of Tanzania since 1997; satisfying demand by the cement and paper industries there (GoM, 1998; 2006; 2008).

The main Malawian consumers of coal are the tobacco, sugar, brewing, ethanol, tea, cement and textile industries, and some public institutions such as hospitals and prisons. Coal has not been adopted for household use because of lack of appropriate domestic cooking appliances, and lack of information, as well as because it is relatively expensive compared to firewood and charcoal (DoE, 2010).

Biomass

Biomass accounts for about 93% of the total primary energy consumption and 99 % of household energy. Biomass is used in the form of firewood 80%, charcoal 8.8%, and agricultural and industrial wastes 11.2%. It is mainly used for cooking, water heating, and space heating. In addition, to meet the bulk of rural and urban people's energy needs, the commercial fuel wood business is a relatively large generator of rural employment, representing some 3.5% of the total GDP (GoM, 1998; 2003).

Biomass is currently used in an unsustainable manner, leading to the destruction of 50,000 to 75,000 hectares of natural forests annually (GoM, 2008). The Government of Malawi estimates household firewood and charcoal consumption at 7.5 million tonnes per annum which exceeds sustainable supply by 3.7 million tonnes. The Government also reports that the demand for forest resources for livelihoods, infrastructure development, and energy are increasing rapidly. Forest degradation is exacerbated by high population growth and poverty thereby stressing the indigenous forests and associated biodiversity. This situation is leading to rapid deforestation, which in turn leads to siltation, sedimentation, flash floods, and the drying-up of streams.

Renewable Energy Technologies

Renewable energy technology installations in Malawi date back to the early 1980s. These were mainly in the form of solar thermal systems for water heating in health centres, and later, solar photovoltaic systems were introduced. By the end of 1999 there were about 5,000 installed systems but only half were operational as of that date. Lack of technical expertise rendered the other half non-functional, sometimes due to only a minor problem.

The most important renewable and alternative energy sources for Malawi are solar, wind, natural gas, biogas and briquettes but contribution towards total energy demand is still very low at less than 1%. Malawi has average wind speeds of between 2 and 6 m/s, good solar radiation resources of around 21 $MJ/m^2/day$, perennial rivers with a potential of 900 MW, large livestock population for biogas generation, and several hot springs for geothermal electricity production. Up to 21 major hot springs are reported in the Chitipa-Karonga area down to Chipudze in the southern region (GoM, 1998; 2003; 2006).

Solar Energy

Solar energy is abundant in Malawi though locally measured data is difficult to find. Internationally available remote sensed databases, however, provide sufficient data for most design purposes. Maximum irradiation of 6.5-7.0 $kWh/day/m^2$ occurs in September-October and the minimum of 4.3-4.6 $kWh/day/m^2$ occurs in January-February or in June-July according to location. The seasonal variation is 30-40 % which can be reduced to 15-20% by selecting the appropriate orientation of the PV panel or the solar collector. The solar energy resource is thus well within the limits normally considered to represent technical and commercial exploitability (GoM, 2006).

4. Case Studies

As of 2008 there were only three villages in Malawi that had a centralised solar-wind hybrid system; three others were under construction. There was one such village system in each of the three regions of the country. The three villages are Kadzuwa Village in the Southern Region, Chigunda Village in Nkhotakota in the Central Region, and Elunyeni Village in Mzimba in the Northern Region. The systems in the three villages were identical. The villages are located in remote areas far from the main urban centres. The Department of Energy worked hand in hand with ESCOM in the planning and installation of these systems. The systems were handed over to ESCOM as part of the Malawi Rural Electrification Programme (MAREP). Having had no prior knowledge of solar and wind technologies, the corporation's engineers attended a training seminar that was arranged for them.

5. Villages

Elunyeni, Kadzuwa and Chigunda wind-solar hybrid systems (see Figure 3) have the following identical specifications: three wind turbines having a power capacity of 4.2 kW each, 55 solar panels with a total capacity of 7.5 kW, one 15 KVA inverter with a single phase 240V output, Battery Bank 240V, 1400 Ah, an output that could support a maximum of 150 houses, each house consisting of up to 5 energy-saver light bulbs (11 watts were recommended) and one power socket (Figure 4 shows electrified houses). The systems also pump water for the village into a 10, 000 litre tank to a network of seven delivery taps, provide street lighting, and provide electricity to a primary school, a dump load consisting of some resistive heaters to cater for excess electricity and a single phase, 240 V local grid.

Figure 3: Solar PV Panels and Wind Turbines

Figure 4: Electrified Houses

6. Health Centres

Three health centres operating under the Christian Health Association of Malawi were visited. These were Kaseye Community Hospital and Sangilo Health Centre in the Northern Region and Nambuma Parish Health Centre in the Central Region. One government-run health centre was visited and this was at Chimaliro in the Southern Region. The health centre in Chimaliro was close to a local court, meaning that the local court could also be easily visited.

7. System Specifications

Solar PV was used to electrify the health centres. The Malawi Bureau of Standards produced a Code of Practice for battery-based solar photovoltaic systems (MS 696:2004). Malawi is at geographic coordinates of 13° 30′ S, 34° 00′ E. For PV modules the following guidelines are given among others: azimuth angle shall be North ±5 º with a tilt angle of 20 ±5 º, avoid ground mounting wherever possible, roof mounting requires 10 cm airflow space from roof, no objects shall shade any part of the panel at least between 8 am and 4 pm.

For batteries, the regulations give the following requirements; the battery should be housed in a box of specified measurements and quality, for institutions the recommended type is the deep cycle, maintenance free battery, the requirement is that the owner should return the dead battery to the installer/supplier for recycling. The code of standard also specifies regulations for inverters, regulators, refrigerators, radio communication systems, wiring procedures, type of lights, and maintenance procedures. The guidelines were prepared in such a way that it is ensured that a system continues working for a period of between three to five years without problems.

The solar PV systems in the health centres were not centralised. Each building had its own individual solar PV installation (see Figure 5).

8. Uses in the House

Some of the staff houses at Nambuma and Chimaliro had solar home systems consisting of a 55 W mono-crystalline solar panel, a 96 Ah, 12 V battery, a regulator, and an inverter. In some cases the inverter was not present.

Figure 5: Solar Electrified House at a Health Centre

Figure 6: Microscope in the laboratory

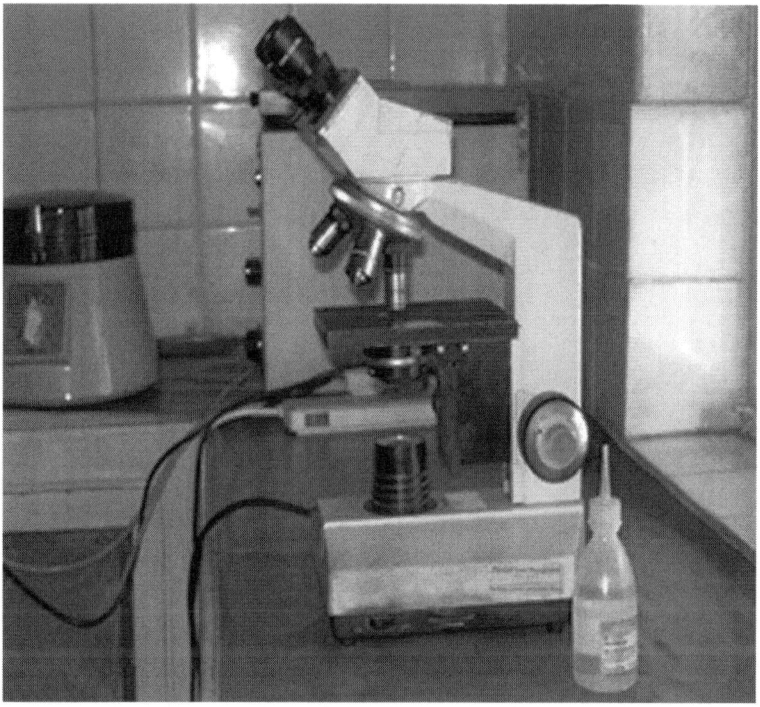

9. Recommendations

The findings from this study show that, if the barriers being faced by Malawi are overcome, renewable energy can be a solution to the rural livelihoods in Malawi. The country has high potential for market growth towards a sustainable energy supply. To achieve this there is need to train barefoot technicians in the rural areas to ensure timely maintenance in case of minor failures. There is a need for extensive research in solar PV and solar thermal. There is also need to investigate solutions to address the lack of finances and poverty, such as subsidies for purchase of SHS by the rural poor and negotiate soft loans for them at the banks or other lending institutions. Feasibility studies for productive use of renewable energy in the rural areas and thereafter implementing them by increasing the projects specifically targeted at productive uses of renewable energy need to be done. There is also a need to increase awareness on the options available to the poor people in the rural areas such as sustainable wood fuel, improved biomass stoves and biogas. It is also essential to ensure that there is availability of readily available spares like lights and inverters close to the villages.

References

GoM, Government of Malawi (1998), *National Sustainable Renewable Energy Programme (NSREP),* Lilongwe.

GoM, Government of Malawi (2003), *National Energy Policy,* Lilongwe.

GoM, Government of Malawi (2008), *Annual Economic Report,* Lilongwe.

GoM, Government of Malawi (2008), *2008 Malawi Millennium Development Goals Report,* Lilongwe.

GoM, Government of Malawi (2006), *Promotion of Alternative Energy Sources Projects,* Lilongwe.

MDHS, Malawi Demographic and Health Survey (2010), *National Report,* Lilongwe.

Tenthani, C., Kaonga, C.C. and Kosamu,I.B.M. (2011), 'The Potential of Distributed Generation in Malawi',in *National Symposium on Climate Change and Environment Conference Proceedings,* Crossroads Hotel, Lilongwe, 19 - 21 October 2011.

UNDP, United Nations Development Programme (2007), 'Energizing Poverty Reduction: A Review of Energy', *Poverty Nexus in Poverty Reduction Strategy Papers,* New York.

Wambua, C. (2011), 'Laws and Policies Enabling the Production of Biofuels in Malawi', *Policy Innovation Systems for Clean Energy Security (PISCES) Working Brief,* 4 May 2011.

Zingano, B.W. (2005), 'The Problem of Fuel Wood Energy Demand in Malawi with Reference to the Construction Industry', Available at: http://www.joyhecht.net/mulanje/refs/Zingano-fuelwood-bricks-2005.pdf.

Acknowledgments

The authors would like to acknowledge Mr. Khumbolawo Lungu for his contributions towards data collection required for preparing this paper.

Chapter 8

Situation of the Research on Renewable Energies in the Central Africa Region: The Case of Solar Energy

César Kapseu[1], Noël Djongyang[2] & Maturin Petsoko[3]

1. Introduction

Multi-lateral institutions and international agreements have acknowledged the link between poverty and energy (e.g. Agence Française de Développement et Banque Africaine de Développement, 2009), the importance of "sustainable energy development" for achieving Millennium Poverty Reduction Goals, and the need for environmental suitability (Showers, 2009). Most of the African countries remain mainly dependent on fossil fuel energies. One of the main fossil fuels, oil production is continuously decreasing. According to some experts, the end of the prevalence of what is called "peak-oil" is expected to occur within 40 to 50 years.

1 Department of Process Engineering, National Advanced School of Agro-Industrial Sciences, University of Ngaoundere, Cameroon. Email: kapseu@yahoo.fr

2 Department of Physics, Faculty of Science, University de Ngaoundere, Cameroon. Email: noeldjongyang@gmail.com

3 FSJP, University of Yaounde 2, Cameroon. Email: maturinpetsoko@yahoo.fr

In addition, these sources of energy emit important amounts of polluting gases that can contribute also to global warming. Renewable energies and particularly solar energy appears to be an ideal alternative to fossil energy (Karekezi and Kithyoma, 2003; Tchuidjan and Nkouimi, 2009). In fact, the sunshine is free, available everywhere, and environmentally clean. It can be used directly (solar thermal) or converted into electricity through photovoltaic (PV) panels. Africa has one of the best climates for this type of energy (IEEF, 2000; Ngnikam, 2006).

The supply and use of energy have never been static subjects. Scientifically, technologies change; some are entirely new and others result in improved function and efficiency. Economically, the primary resource base changes, with some resources being indigenous and therefore relatively secure, while other resources are imported with significantly less security (UN Energy/Africa, 2008). Central African countries depend for about 95 % on hydropower. That region has the greatest African hydraulic potential with the Democratic Republic of Congo and Cameroon respectively first and second. Grand Inga on Congo River was an investor's dream for most of the twentieth century. However, common sense engineering appraisal revealed catastrophic terrestrial consequences from such a large project (Showers, 2009). In addition, during the dry seasons, the level of the water in the dam is generally low and consequently leads to electricity shortage with negative consequences on local economy. A solution could be to support the development of the renewable energy sector which is stable and environmentally clean. The prospects offered by renewable energies are promising: fight against desertification, limitation of the impacts of climate change, and development of large scale "decentralized electrification" (Agence Française de Développement et Banque Africaine de Développement, 2009). The resort to these energies would involve a better access of the populations to the energy services, the reduction of poverty by creating incomes and protecting the environment.

This study aims to present an overview of the current status of the research in solar energy in the central Africa region. Some recommendations will also be provided to boost the researches in the field of the renewable energies as well as their implementation.

2. The Central Africa Region

The central Africa region (Figure 1) is the core area of the African continent. It includes 11 States: Angola, Burundi, Cameroon, Gabon, Equatorial Guinea, Central African Republic, Democratic Republic of Congo, Republic of Congo, Rwanda, Sao Tome and Principe, and Chad (United Nations, 2000; Africa-Union, 2007; Encyclopedia Britannica, 2007). The total area of the region is estimated as 6 707 889 km^2, while its population is estimated to 115 130 510 Inhabitants (Groupe Vinado Fandio, 2010).

With a mean global solar radiation of 1 788 kWh/year/m^2(Nkué and Njomo, 2009 and Ayangma *et al.*, 2008), the Central Africa region represents a natural solar energy potential of about 12 GWh/year as presented in Table 1, but that potential is currently largely unexploited.

Two main types of climate patterns prevail: wet-tropical and dry-tropical. The vegetation is made up of savannah (grasses and shrubs) in the dry-tropical region and forest galleries (big trees) in the wet-tropical region.

Figure 1: Central Africa

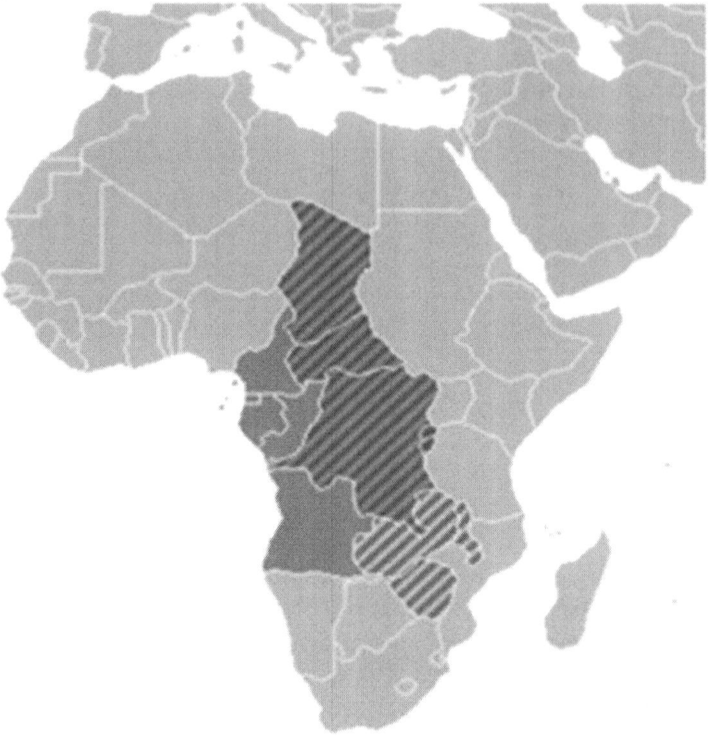

Source: Website A

Table 1: Mean Solar Energy Potential, Central Africa Region

Country	Area[ϒ] (km²)	Population (inhabitants)	Average solar energy* potentiality (GWh/year)
Angola	1246700	12356940	2.229
Burundi	27834	6677950	0.050
Cameroon	475442	19406100	0.850
Congo	342442	3500000	0.612
Gabon	268000	1208410	0.479
Equatorial Guinea	28051	496000	0.050
Central African Republic	662984	3539810	1.185
Democratic Republic of Congo	2345095	50500000	4.193
Rwanda	26340	8310000	0.047
Sao Tome and Principe	1001	145300	0.002
Chad	1284000	8990000	2.296
Total	6707889	115130510	11.994

Notes: ϒ- Groupe Vinado Fandio (2010); *-Estimated from the average value of the global solar radiation of 1788 kWh/year/km² (Nkué and Njomo, 2009; Ayangma *et al.*, 2008)

3. Current Status of the Research in Solar Energy

Solar energy is mostly used on a large scale in the central Africa region under its direct form (drying processes, water heating, etc.). However, its indirect use (electricity through PV panels, solar pumping, etc.) is growing rapidly. Despite some lateness compared to the western or eastern Africa regions, the research in the field of solar energy in the central Africa region is promising. Engineering schools, private institutes, programs facilitating the expansion of the use of solar energy, and nongovernmental organizations (NGOs) have embarked on research to produce technologies working with the sun. Sales points and service structures devoted to solar technologies are also increasing. Many universities have laboratories working in the field of solar energy; even if most of them are mainly theoretical due to the fact that the research equipment and test installations are too expensive for their budget. The central African countries rely heavily on agriculture, solar pumping would be highly useful. Some applications of these technologies are already used in the northern part of Cameroon as well as other countries like Chad.

Table 2 presents some central African's research laboratories working in the field of the renewable energies. The research units are generally organized around a coordinator who is more often a lecturer or a technician.

Table 2: Some Research Laboratories in Solar Energy in the Central Africa Region

Country	Affiliation	Laboratories
Cameroon	University of Ngaoundere	Reaction, Extraction, Sugar refinery, Drying, Oil mill
		Laboratory of Energy
	University of Maroua (Higher Institute of Sahel)	Renewable Energies
	University of Yaounde I (Faculty of Science)	Environmental Energy Technologies Laboratory (EETL)
	University of Yaounde I (National Advanced Polytechnic School)	Laboratory of Energy
	University of Dschang (Fotso Victor University Institute of Technology, Bandjoun)	Laboratory of Industrial Systems Engineering and Environment, (LISEE)
	University of Douala (University Institute of Technology)	Thermal Engineering

Republic of Congo	Marien Ngouabi University of Brazzaville	Mechanical and electrical engineering
Chad	University of N'djamena	Applied Physics
Gabon	University of Science and Technologies of Masuku (Franceville)	Science and Technologies
Democratic Republic of Congo	University of Kinshasa	Laboratory of Biotechnology and Environment
Central African Republic	University of Bangui	Carnot's Laboratory of Energy
Equatorial Guinea	National University of Equatorial Guinea	Escuela universitaria de Ingeniarias tecnicas EUIT (BATA)
Burundi	University of Burundi	Laboratory of New and Renewable Energies
Rwanda	University of Rwanda	Laboratory of Applied Energy
Angola	University of Luanda	Architecture

Various topics are developed in these laboratories as presented in Table 3. Some results are published in journals or congress proceedings (e.g. World Renewable Energy Congress XI, Abu Dhabi, 2010; First National Symposium on Energetic Systems, 2004, Bandjoun, Cameroon). However, it appears largely that the results are under the expectations of the local populations. To encourage research in the field of solar energy, it would be necessary to encourage projects of regional integration through international organizations like ANSOLE, especially devoted to bringing together all research activities over the continent. Cameroon hosted the ANSOLE Days in February 2012. Such meetings are to be encouraged in order to create networks and collaborations among researchers so that to develop the sector of solar energy.

It appears research in the field of solar energy is growing, though application and implementation have not taken off yet for a number of reasons that are addressed in the next section.

Table 3: Distribution of Research Topics in Solar Energy

Country	City	Research topics
Cameroon	Ngaoundere	Solar drying, Renewable energies
	Maroua	Solar drying, Renewable energies, Installation and maintenance of solar equipment
	Bandjoun	Solar thermal, Solar Photovoltaic, Solar concentration
	Yaounde (Faculty of Science)	Drying processes, Solar thermal, Solar Photovoltaic
	Yaounde (National Advanced Polytechnic school)	Solar drying, Renewable energies
Congo	Brazzaville	Solar energy, Energy efficiency, Installation and maintenance of the solar equipment
Chad	Ndjamena	Solar drying, Renewable energies
Gabon	Franceville	Process Engineering, Renewable Energy
Democratic Republic of Congo	Kinshasa	Renewable energies, Biofuels
Central African Republic	Bangui	Energy systems
Equatorial Guinea	Bata	Energy and Indoor air Quality

Burundi	Bujumbura	Solar thermal and photovoltaic
Rwanda	Kigali	Renewable energies
Angola	Luanda	Indoor Air Quality, Oils, Renewable energies

4. Some Obstacles to the Implementation of the Renewable Energy Sector

Few investments are made in the field of the renewable energies in the central Africa region. Obstacles to the development of that important field of activity are political, technical, institutional and financial. In fact, the slowness of the political and institutional reforms necessary to the creation of a favourable environment to private or foreign investments in the energy sector constituted indeed one of the major obstacles. Few attentions are also paid to access energy technologies, reduction of social inequalities, environmental protection, mobilization of financial resources, and reinforcement of institutional and human capacities. In addition, one can enumerate: high importation taxes for PV panels (custom charges); lack of suitable policies and information on renewable energy's new technologies; absence of a framework allowing and encouraging the participation of the private sector into the energy production; lack of qualified personnel, coordination and sharing experience in a regional and continental scale; etc. However, the main problem for many countries for having energy on a large scale is import taxes, local corruption, theft, availability of spare parts and willingness of local installers for installing a solar module on a real remote location.

5. Some Recommendations

Despite the great potential in solar energy, this form of energy is less supported by local authorities. However, the will of rural and urban households to benefit from this form of energy is acknowledged by several African Governments. Despite the advantages offered by solar energy, the cost of PV systems is still exorbitant. Yet, more homes could benefit from solar energy, if the governments suppress the import taxes. The use of solar panels supplying several homes at once can contribute to cheap electricity. Development of solar markets with special funds to facilitate consumer credit as done in most of the countries of southern Africa could also be a solution. In addition, the authorities should:

- Develop, encourage and establish close cooperation among the financial, institutional and governmental partners to support renewable energy projects;
- Regularly organize seminars on renewable energies between the researchers, partners of the private sectors, and the institutions;
- Conceive, finance and implement solar projects in each country with sub-regional or regional extensions;
- Maintain a dialogue between producers and consumers of solar energy in order to increase transparency and exchange information;
- Re-examine energy policies to support an accelerated development and application of high-output solar technologies;
- Intensify international cooperation, particularly through North-South and South-South exchanges. It is advisable to be interested in international cooperation supporting local development, adaptation, exploitation and safeguarding less polluting technologies. A particular attention should also be paid to the reinforcement of the capacities which will make possible to locally produce equipment necessary to ensure a sustainable energy provisioning;
- Intensify the marketing and promotion actions of new solar technologies as well as increasing the scientific and technological cooperation by the means of researches, development, transfer and diffusion of respectful technologies for environment;

- Produce by making and encouraging political, legal and institutional frameworks targeted to renewable energies;
- Set up long term policies and programs on renewable energies, including the private sector;
- Liberalize the energy sector and develop local energy technologies;
- Encourage projects on renewable energy, especially sub-regional projects. An essential recommendation is to associate the NGOs and the civil companies into the realization of these projects.
- Reorganize and create special financing structures for Renewable Energy Control Agencies.

6. Conclusion

The current status of the research in solar energy in the central Africa region has been explored. It is encouraging that the research increasingly growing. Public and private schools, programs facilitating the expansion of the use of solar energy, and nongovernmental organizations have embarked on research to produce technologies working with the Sun. Sales points and service structures devoted to solar technologies are also increasing. Many Universities have laboratories working in the field of solar energy; even if most of them are mainly theoretical due to the fact that the research equipment and test installations are too expensive for their budget. However, even if research in the field of solar energy is promising, its implementation is still to take off due to political, technical, institutional or financial constraints that affect both research and application. Real progress in both research and implementation has begun in central Africa. Examination of the ongoing projects demonstrates that the sun has a future in the Central Africa and the relevant stakeholders are increasingly aware of the potentiality of solar energy to fulfil renewable energy for the region.

References

Africa-Union (2007), 'Economic Community of Central African States', Available at: http://www.africa-union.org/root/au/recs/eccas.htm (Accessed: 10 February 2012).

Agence Française de Développement et Banque Africaine de Développement (2009), 'L'Energie en Afrique à l'horizon 2050', *Report*.

Ayangma, F., Nkeng, G. E., Bonoma, D.B. and Nganhou, J. (2008), 'Evaluation du potentiel solaire au Cameroun: cas du Nord Cameroun', in *African Journal of Science and Technology (AJST) Science and Engineering* series 9(2): 32-40.

Encyclopaedia Britannica (2007), 'The Central African Federation', Available at: http://www.britannica.com/eb/article-43831/Southern-Africa(Accessed: 10 February 2012).

Groupe Vinado Fandio, (2010), 'Le monde et ses six continents', Edition revue, corrigée, actualisée, Douala, Cameroun.

IEEF, Institut de l'Energie et de l'Environnement de la Francophonie (2000), 'Le Programme solaire mondial une nouvelle dynamique pour les EnR', in *Liaison-Energie-Francophonie* N°46 – 1er trimestre 2000.

Karekezi, S. and Kithyoma, W. (2003), 'Renewable energy development', in *The Workshop for African Energy Experts on Operationalizing the NEPAD Energy Initiative* 2- 4 June, 2003 Novotel, Dakar, Senegal.

Nkué, V. and Njomo, D. (2009), 'Analyse du système énergétique camerounais dans une perspective de développement soutenable', in *Revue de l'Energie* vol. 588, Mars-Avril 2009.

Ngnikam, E. (2006), 'Énergie et écodéveloppement au Cameroun', *Helio International*, Observatoire de la viabilité énergétique 2005/2006.

United Nations (2000), 'World Macro Regions and Components', Available at: http://www.un.org/depts/dhl/maplib/worldregions.htm(Accessed: 10 February 2012).

UN-ENERGY/Africa (2008), 'Energy for Sustainable Development: Policy Options for Africa', *UN-ENERGY/Africa publication to CSD*$_{15}$.

Showers, K.B. (2009), 'Congo River's Grand Inga hydroelectricity scheme: linking environmental history', policy and impact, in *Water Hist*, 1: 31–58.

Tchuidjan, R. and Tchana Nkouimi, M.J. (2009), 'Outil d'aide au dimensionnement des systèmes photovoltaïques domestiques', in *Revue des Energies Renouvelables*, 12(3): 523–532.

Website A: http://www.wikipedia.org (Accessed: 10 February 2012).

Chapter 9

Investigation of the Solar Energy Production and Contribution in South Africa

Silas K. Mulaudzi[1], Mammo Muchie[2] & Rudzani Makhado[3]

1. Introduction

Globally, the renewable energy industry is projected to grow rapidly in the next few years (REN21, 2009). The International Energy Agency estimated that 15-20% of the total energy supply contribution will be generated from renewable energy by the year 2026. Investment in renewable energy sources for electricity, heating and in biofuels has increased considerably. Recent statistics shows that nearly 11 GW of solar PV was produced globally (REN21, 2009; REN21, 2010). According to REN21 (2010) first Solar, which is in the USA, became the first firm ever to produce over 1

1 Energy and Electricity Division, City of Tshwane, P.O Box 440, Pretoria, 0001, South Africa. Email: Silasmu@tshwane.gov.za

2 DST/NRF SARChI Professor (Innovation & Development), Tshwane University of Technology, South Africa; Technology and Management Centre for Development (TMCD), Oxford University, UK; Associate Faculty Professor, Sussex University, UK; Adama Science and Technology University (ASTU), Ethiopia. Email: MuchieM@tut.ac.za

3 Research and Evaluation Section, Limpopo Legislature, P/Bag x9309, Polokwane, 0700, South Africa. Email: makhador@limpopoleg.gov.za

GW in a single year. Major crystalline module price declines took place, by 50–60% by some estimates, from highs of $3.50 per watt in 2008 to lows approaching $2 per watt (Chinese Renewable Energy status, 2009).

Energy utilisation in South Africa is characterized by high dependence on cheap and abundantly available coal. South Africa also imports a large amount of crude oil; but a limited quantity of natural gas is also available (Digest of South African Energy Statistics, 2009). South Africa has among some of the best solar resources in the world, the country has already committed itself to a target of 10,000 GWh of renewable energy by 2013 (DME, 2003). According to the Long-Term Mitigation Strategy achieving renewable electricity supply targets of at least 27% by 2030 and 50% by 2050 would require a major rollout of Concentrated Solar Power (CSP) and solar Photovoltaic (PV) generation capacity. CSP is projected to contribute an installed capacity of up to 150 GW and solar PV up to 50 GW by 2050 (Winkler, 2007).

Solar energy is categorized into concentrated solar power, solar photovoltaic and solar water heating. South Africa is regarded as having good solar potential in areas around Limpopo, Northern Cape, North West and Mpumalanga Provinces (Figure 1).

Figure 1: Solar Energy Potential in South Africa.

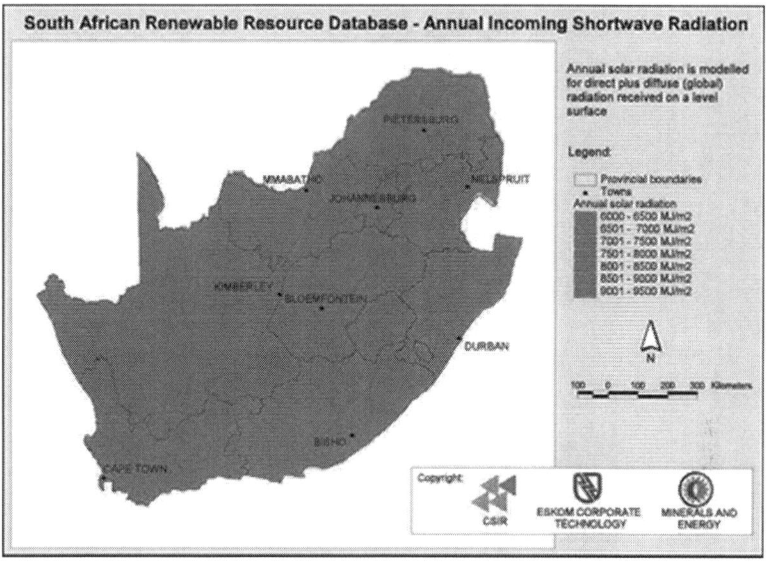

Source: DME et *al.*, (2001)

Solar radiation levels in South Africa are amongst the highest in the world. Average daily solar radiation varies between 4.5 and 7 KWh/m2 (Banks and Schaffler, 2005). South Africa has the potential of generating up to 24 000 MW of CSP by 2030 given the excellent solar radiation available and infrastructure to support a project of such a magnitude (Edkins *et al.*, 2009).

The concentrated solar power is the largest existing centralized system that uses parabolic troughs, which focus sunlight into evacuated glass tubes carrying the heat to conventional steam turbines via heat exchangers (see Table 1). Solar photovoltaic cells make up modules that are placed in arrays and convert sunlight into direct current electricity without any moving parts (New Energy Finance, 2009).

Table 1: Solar Energy Technologies	
Solar Photovoltaic	**Concentrated Solar Power**
• It generates electrical power by converting solar radiation into direct current electricity using semiconductor	• It uses mirrors or lenses to concentrate a large area of sunlight or solar thermal energy onto a small area
• Material used for photovoltaic are monocrystalline silicon, polycrystalline silicon, copper indium gallium selenide	• Electrical power is produced when concentrated light is converted to heat which drives heat engine connected to an electrical power generator

Source: REN21 (2009); Environment America (2008).

2. The Solar Energy Development

CSP Development

Globally, CSP has four major technology designs, with a total of 430 MW in service and more than 8.7 GW capacity planned (Table 2).

- Parabolic troughs are systems whereby special mirrors shaped as linear parabolas reflect the sun's rays toward an absorption tube suspended at the centre of the trough's arc. The concentrated sunlight heats fluid inside the tube, generally a synthetic oil up to 400°C, which then travels to a collecting unit, where it heats water and generates steam to power turbines (Environment America, 2008). The troughs are typically arrayed on a north-south axis and track the sun throughout the day. The Solar Thermal Energy Generation (STEG) plants in the US demonstrate the longest track record of this technology (17 years), which continues to gain support today, with 395 MW of capacity in service and almost an additional 5 000 MW capacity in planning (Environment America, 2008).

- Concentrating dish/Stirling engines are shaped like a satellite dish. They focus light rays on a single area suspended above the bowl of mirrors, where temperatures up to 750°C heat a thermal fluid, which in turn runs a small steam or Stirling engine (Mills and Morrison, 1997). Although they have in the past been considered more useful as independent, off-grid units, particularly in remote and developing areas, the recent developments announced by Stirling Energy Systems, who plan to build up to 1.75 GW of solar farms in the California desert, indicate their additional attractiveness for connection to the national grid. Unlike the other CSP technologies, Stirling engines do not need water for cooling, though water is still necessary to clean the mirrors (Environment America, 2008).
- Central receivers/towers use freestanding heliostats (tracking mirrors) in an array to independently track the sun and focus its rays onto a central tower. Temperatures of up to 650°C are achieved, they require less land than parabolic troughs, and because the heliostats used as reflectors are nearly flat, their manufacturing costs are relatively low (Environment America, 2008). Demonstration central receiver plants have been built, beginning with a 0.5 MW test plant in 1981 in Spain and followed by others in France, Italy, Japan, Russia and the USA (Environment America, 2008). The first commercial plant, an 11 MW central receiver plant developed by Abengoa Solar, went online in March 2007 to deliver power to the city of Seville in Spain. Abengoa has since built a second 22 MW plant, with an additional 600 MW capacity in the planning process (REN21, 2009).
- Linear Fresnel reflectors (LFRs) use long rows of nearly flat, rotating mirrors to reflect light at absorbers elevated above the plane of the mirrors. Different absorbers use either a thermal transfer fluid or directly generate steam to power turbines. While not as efficient as parabolic dishes and troughs or central receivers, LFRs offer many

potential cost and structural advantages. Like central receivers, their mirrors are made of standard glass in large, flat sheets, which require fewer steel supports than parabolic troughs and can be cheaply mass-produced (Environment America, 2008).

The mirrors' flat-shape renders them more resistant to wind damage and makes them easier to clean (Clean Energy Action, 2007). LFRs' fixed absorbers do not have moving joints, which simplify fabrication and avoid the cost and maintenance challenges presented by joints in parabolic trough arrays, and the recent development of the Compact Linear Fresnel Reflector (CLFR) has removed the problem of shadows (Mills and Morrison, 1997). Demonstration LFR arrays have been built in Australia, Spain and Belgium and in the US (Environment America, 2008). Originally conceived in the early 1990s by Sydney University, CLFR was first commercialised in 2004 in Australia and is now being refined and built on a large scale, with about 1500 MW of capacity in the planning (Environment America, 2008).

Table 2: Summary of CSP Development in 2008

Technology	In Service Capacity (MW)	Planned Capacity (MW)	Total	Leading Locations (Including Planned Installation)
Trough	395	4 967	5362	US, Spain, China, Israel, Australia, Morocco, Greece, UAE, Algeria, India, Mexico, Iran
CLFR	1	1 489	1490	US, Libya
Tower	33	579	612	Spain, US, South Africa, Egypt
Dish	1	1 750	1751	US
Total	430	8 785	9215	

Source: Staley *et al.*, (2009); Spellmann (2009).

Solar PV Development

Solar PV generates electricity in well over 100 countries and continues to be the fastest growing power-generation technology in the world. Between 2004 and 2009, grid-connected PV capacity increased at an annual average rate of 60 percent (REN21, 2010). In 2010, solar PV installed capacity grew up to 40 MW (Figure 2). Germany again became the primary driver of PV installations, more than making up for the Spanish gap with 3.8 GW added—about 54 percent of the global market. Germany installed a total of 47% of total installed GW capacity globally (Figure 3).

The United States added an estimated 470 MW of solar PV in 2009, including 40 MW of off-grid PV, bringing cumulative capacity above the 1 GW mark with 6% of the total global PV installed (Figure 3).

The crystalline silicon solar cell remains the mainstream of commercialized solar cell nowadays. In international market, over 90% of solar cells are made of high purity polysilicon (Chinese Renewable Energy Status Report, 2009). Polysilicon is material consisting of small silicon crystal that differs from single-crystal silicon and is used for electronics and solar cells, and from amorphous silicon, used for thin film devices and solar cells.

Figure 2: Solar Photovoltaic, Existing World Capacity, 2000-2010, Source: REN21 (2010)

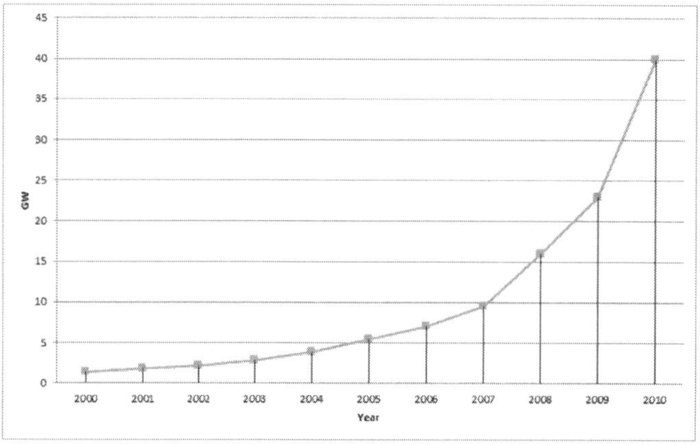

Figure 3: Solar PV existing capacity, 2009 (REN21, 2010)

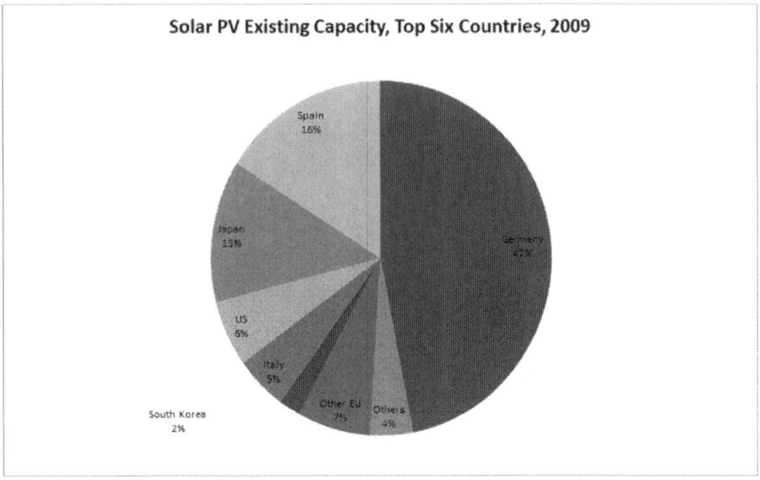

As the most fundamental materials in PV cell, the high purity polysilicon has become the most significant part in PV industry chain (Chinese Renewable Energy Status Report, 2009). This is the biggest barrier for the PV industry in China. From 2001, the Chinese Polysilicon Production has increased drastically (Table 3). The increase was largely due to reduction in polysilicon price. China's solar PV industry is also growing rapidly and the country became the world's largest PV cells producer by 2008 (REN21, 2009).

Table 3: China Polysilicon Production, 2001◎2008 (tons)	
Year	**Production(tons)**
2001	40
2002	50
2003	60
2004	70
2005	80
2006	290
2007	1140
2008	5500

Source: Chinese Renewable Energy Status Report (2009)

3. Solar Energy Status in South Africa

South Africa has among some of the best solar resources in the world, the country has already committed itself to a target of 10 000GWh of renewable energy by 2013 (DME, 2003). Moreover, the country has developed the Integrated Resource Plan (IRP) that has a committed total of 600 MW of solar PV by 2019 and 300 MW of CSP by 2019 (DoE, 2011).

South Africa has various types of renewable energy technologies that are manufactured locally and internationally. Most of the wind and hydro generators are locally manufactured (Table 4). Some of the manufactured hydro, wind and solar Photovoltaic (PV) are exported regionally and across the continent. However, large wind generators and turbines are imported from the American and European countries.

The government of South Africa has established a mini-grid hybrid system in the Eastern Cape Province in 2004. The Lucingweni Hybrid mini-grid system consisted of power generation, reticulation and premises. The power generation consists of wind turbines, using 6x6 kW Proven Wind Turbine mounted on 6m tall mast and 560 Shell Solar 100W Panels mounted on steel structure and an equipment cabin/shelter constructed on a concrete base (DME, 2006). The total solar PV capacity was 56 000 watts (56 kW).

The hybrid system supplied power to approximately 220 households in Lucingweni area in the Eastern Cape for more than 6 months. According to the DME (2006) the system worked well until the illegal connection started and put much pressure on the supply. Moreover, the power was supposed to be limited to 2 Amps and it was mistakenly connected at 20 Amps, which resulted in situation where one household could withdraw all power at a specific time during the day. The system was vandalised by the local community and eventually it could not generate any electricity (DME, 2006).

Table 4: Renewable Energy Manufacturing Companies in South Africa

Company	Business type	Technology	Location
Solar Deluxe	Manufacturers, wholesale supplier	Solar water heating	Cape Town, Western Cape
Budget Solar (Pty) Ltd	Manufacturers/wholesale/ importer	Solar electric power	Somerset West, Western Cape
Pre-plan energy	Installer/developers	Wind, biogas, Solar Photovoltaic	Alberton, Gauteng
Solartech	Manufacturers and installers	Solar water heating	Lonehill, Gauteng
Solar heat exchanger	Manufacturers and installers	Solar water heating	Johannesburg, Gauteng
Solar Beam	Manufacturers/installers	Solar water heating	Rivonia, Gauteng
Environment Save Energy	Manufacturer	Wind turbine and solar photovoltaic	Port Elizabeth, Eastern Cape
Green Wind Power and automation	Manufacturer/wholesale supplier	Wind turbine	Cape Town, Western Cape
Solardome SA cc	Design and installation	Solar water heating, solar thermal storage	Stellenbosch, Western Cape
Renergy Technologies	Manufacturer, exporter, importer	solar panel, solar batteries, solar water heating	Elandsfontein, Gauteng
All power trust	Manufacturers/ importers/ exporters	Solar electric power	Port Elizabeth, Eastern Cape
Inti Solar	Manufacturers and installers	Solar water heating	Fourways, Gauteng

Source: See: http://energy.sourceguides.com/business/byGeo/byC/SouthAfrica

4. Objectives of the study

The objectives of this study are:

- To establish the import and export relationship between the domestic and international firms that manufacture solar energy technologies;
- To investigate the production and contribution of energy in South Africa's energy growth; and
- To determine the impediments hampering the growth of solar energy manufacturing industry in South Africa.

5. Methodology

The study was conducted in South Africa, thus covering nine Provinces of the country. However, electricity consumption data was collected in five Provinces, namely: Gauteng, Mpumalanga, Free State, Northern Cape and North West (see Figure 4). Other data was collected through the distribution of questionnaire in various organisations in Provinces across the country.

Figure 4: Study Area for the Research (Location where the Study was conducted)

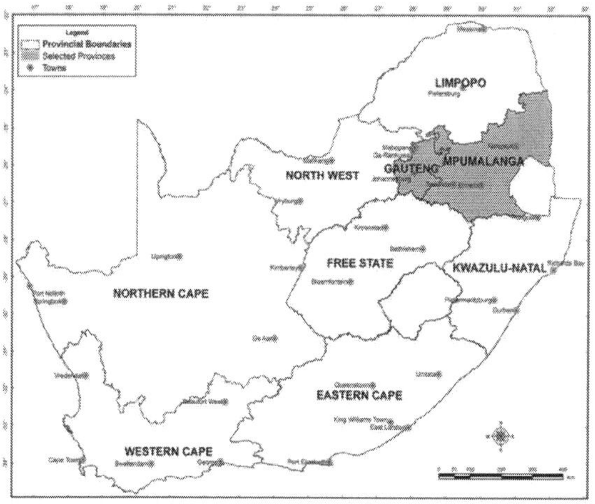

The mixed (qualitative and quantitative) technique for data collection and analysis was used in this paper. The primary data was collected randomly through interviewing the respondents. A questionnaire was used to capture the collected data and information. Secondary data was collected from various sources including books, newspaper, journals, television and government statistics.

Primary data was collected from the randomly selecting the following sectors as shown in Table 5.

Table 5: Sources of Primary Data

Sector	Organization
Manufacturing	• Inti solar Company • Okhela renewable energy corporation • Concentrix solar
Academic institutions	• One respondent from University of Johannesburg • One respondent from Fort Hare Institute of Technology • One respondent from University of Cape Town • One respondent from Wits University
Consumers	• 10 respondents from Secunda • 10 respondents from Tshwane • 5 respondents from Bloemfontein • 2 respondents from Kimberly • 2 respondents from Johannesburg
Generators/Producers	• 3 respondents from Eskom

The level of investigation was sampling. The data collected from sampling gave a good indication of the measurement, facts and other parameters of the entire population from which samples are drawn. In this paper sampling was used because it involves less cost, less amount of time to complete and is more practicable.

A stratified random sampling was used to select the sample in order to give all individual in a population equal chance for being included in the sample and is free of biasness. It was used because the data is required from two or more different types of attributes such as government, Non-governmental organizations (NGOs) or private organization. People that were sampled for interviews were from Industry Associations representatives, State Owned Entities (SoE), individuals from the households, private sectors that deal with energy issues in South Africa.

The collected data was captured and stored in MS Excel 2003 software. Data was analysed both quantitatively and qualitatively. Data was analysed quantitatively using descriptive statistics, analysis of variance.

6. Results and Discussion

Import and Export Relationship between the Domestic and International Firms that Manufacture Solar Energy Technologies

This study found that many solar PV firms in South Africa are subsidiaries of European and North Americans' firms. It was found that 80% of the firms were established as subdivisions of the European firms. The majority of firms at this point in time are highly involved in marketing the products to the commercial and industrial customers in the country whilst waiting for the Government of South Africa to finalise the feed-in tariff scheme. It was found that 5 commercial and industrial companies have installed solar PV for own use power generation.

This study further found that the raw materials are exported to the parent firms in Europe and North America. While some quantities of Solar PV cells are manufactured in South Africa, majority are manufactured outside the country. It was found that South Africa import the most solar PV from Europe, Asia and America. The reasons given are that it is cheaper and cost effective to transport raw materials than equipment. Moreover, the raw materials will enable the firms to create more jobs and skills development in South Africa.

The DME established Lucingweni mini-grid hybrid system in the Eastern Cape, which was a combination of wind and solar PV modules. The solar PV modules had a total capacity of 56 kW. The PV cells were imported from UK and Germany. These substantiate the fact of export and import of raw materials and final product from and to South Africa.

We found that there's a substantial interest from the Europeans and Americans intending to establish CSP in South Africa. The Northern Cape Province is targeted for the establishment of the CSP. The Government of South Africa has signed Memorandum of Understandings (MoU) with many Asian and European countries on solar energy development. It transpired that the technology advancement moves at faster rate than the policy and regulatory framework of South Africa. That is if the government had finalised the subsidy schemes including feed-in tariff the country would have substantial amount of electricity generated from CSP and solar PV and the industrial development including manufacturing sector for solar would be highly advanced.

Preference of Energy Sources (Per Province)

In this study the residents in two Provinces (Gauteng and Mpumalanga) were interviewed in this paper. The purpose was to determine which source of energy between nuclear, coal and solar would residents prefer given a choice. The respondents were middle class and educated. The questions asked were what source of electricity would they like to consume and why do they prefer to consume electricity generated from such sources?

Figure 5: Preferred Source of Electricity in the Mpumalanga and Gauteng Provinces

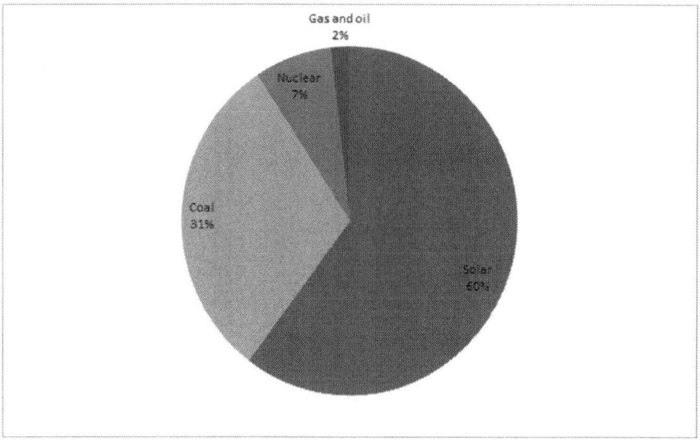

In both Provinces, approximately 60% of the respondents chose solar energy as the preferred source of energy. The reasons cited were that it is natural and has no negative effect such as greenhouse gas emissions. About 31% chose coal because it is abundant in South Africa and has created many jobs in the mining sector, while 7% chose nuclear because it is reliable and has no side effects such as greenhouse gas emissions and would be best if the health hazards risks are managed well. The remaining 2% of the respondents chose to use gas and oil because they can be extracted for multiple uses such as petrol for vehicle, gas for cooking and electricity (see Figure 5).

Table 6 shows the preferred source of electricity between Mpumalanga and Gauteng provinces. About 75% of respondents in the Mpumalanga Province prefer to use solar, while 55% in the Gauteng also prefer to use the same source of electricity. Ten percent of respondents in Gauteng indicated that they prefer to use electricity generated from coal, whilst 45% in Mpumalanga prefers to use coal as source of electricity. The highest percentage (45%) in Mpumalanga prefers to use coal for electricity generation because coal is more abundant, and that coal mining contribute significantly in the GDP of the province as well as job creation and economic development. Twelve percent of respondents in Gauteng and 4% in Mpumalanga prefers to use nuclear as source of electricity. The use of gas and oil only accounted for 3% in Gauteng and 1% in the Mpumalanga provinces (see Table 6).

Table 6: Preferred Electricity Source in Gauteng and Mpumalanga

Source	Gauteng (%)	Mpumalanga (%)
Solar	75	55
Coal	10	45
Nuclear	12	4
Gas and Oil	3	1

Growth Impediments in the Solar Energy Manufacturing Industry in South Africa

South Africa is well-endowed with natural resources that are abundant. It is clear that renewable energy development will require financial incentives. While the Government intends to provide the necessary incentives, South Africa's fiscal resources are limited, and there are competing high priorities social and economic programs, particularly in providing services to historically disadvantaged communities (DME, 2003).

The academic, research, Government and association institutions were interviewed about the growth impediments in the industry and it was found that there are major factor hampering the development of solar energy in South Africa. The barriers vary from the institutional arrangement, regulatory framework, capacity building, training and skills, policy uncertainty and gaps as well as research and development.

Regulatory Framework

It was found that slow pace of finalising feed-in tariff and other related regulations to support renewable energy in general and solar PV and CSP in particular is a critical stumbling block that is hampering the development of solar energy in South Africa. It was acknowledged that the introduction of renewable energy tariffs by the National Energy Regulator was a step in the right direction.

It was further acknowledged that introduction of the Independent Power Producer (IPP) programme has been a breakthrough and the expression of interest advertised where the IPP could apply was a great progress to the realisation of solar energy in South Africa.

According to DoE (2011), the firm commitment necessary for now requires 1200 MW of solar PV at 300 MW from 2012 to 2015 and 100 MW for CSP in 2016. The final commitment of a total of 1200 MW from solar PV starting from 2016 at 300 MW until 2019 (DoE, 2011). The CSP has final commitment of a total 300 MW for 3 years at 100 MW each year from 2017 to 2019 (DoE, 2011).

The White Paper on Renewable Energy set a target of 10 000 GWh by 2013. It was an important move towards promotion of renewable energy in general and solar energy in particular. The White Paper's target was reviewed in 2009. According to the Integrated Resource Plan (DoE, 2011), which has published the energy carriers for 2010 – 2030, there is a good solar PV and CSP allocation from 2016 to 2019 committed. This was seen as a good positive move towards the introduction of renewable energy and diversification of energy resources in South Africa.

Capacity Building, Training and Skills, and Research and Development

It was found that there is a great need for capacity building, training and skills in the solar energy sector. The country's Further Education Training (FET) should be geared to training students and enhance skills for sector specific. According to the research institution interviewed there is no sufficient solar energy training and skills enhancement programme in most of the FET colleges. The academic institutions indicated that the established hubs for renewable energy such as the Stellenbosch University should be equipped with all resources including sufficient funding to accommodate more students.

With the development of PV industry, the related training activities, organizations and mechanism begin their development as well. But with the prediction of explosive growth of PV industry in the coming 15 years, in order to satisfy the huge training demands from vocational education, university education even public awareness dissemination, a comprehensive, sustainable and capable training system is important (Chinese Renewable Energy Status report, 2009).

According to the International Energy Agency (IEA, 2008), government energy research and development budgets in many countries have declined between early 1980s and the 1990s from USD 18 billion to 8 billion in 1997. Where the research and development investments are made, returns can be high; the average return on research and development firms I s estimated to around 20% - 30% (IEA, 2008).

It was emphasised that the government should invest significantly in the research and development through the relevant institutions such as SANEDI, CSIR and the Department of Science and Technology. The research and development institutions should be empowered to establish a comprehensive programme on researching and developing CSP and PV technologies that could gain from South Africa's present industry-base, which are aimed at encouraging domestic CSP component production, CSP technologies that allow for greater storage abilities and technologies that result in less water use, such as dry cooling or using ocean water for cooling. The international funding such as Renewable Energy Market Transformation is supporting renewable energy sector including solar energy technology, however, further CSP research and development, as well as diffusion, need to take place (Edskin *et al.*, 2009).

The China Renewable Energy status report (2009) shared the same sentiment that it is meaningful to establish national solar PV and CSP research and development centre for enhancing the core technology and comprehensive technology indicators of PV and CSP industry. The establishment of a sustainable solar PV industry supporting system includes fundamental research and development, key technology, industrialized equipment production, public research and testing accreditation platform (Chinese Renewable Energy Status Report, 2009).

7. Recommendations

The results of this research have both negative and positive impacts in the development and production of solar PV and CSP in South Africa. There is therefore a scope for improvement from all role players in the industry. To this end, the following recommendations are made:

Regulatory framework

The government should structure the feed-in tariff and the IPP programme in such a way and manner that it will accommodate small players also, not just the big firms.

It is also recommended that the IRP increase the solar PV and CSP allocation in the final report. It is believed that (Figure 1) there is a huge solar energy potential which needs to be harnessed, and therefore the 600 solar PV and 300 CSP allocated by 2019 is insufficient.

Capacity building, training, skills and awareness

This paper recommends that FET colleges and the established hubs across the country be empowered to offer comprehensive training on solar energy technologies. There should be a proper coordination whereby the graduates from the FET colleges are sent to the training centres whereby they be offered experiential learning.

There should be a policy on skills retention to ensure that the trained personnel in the country should serve in the domestic firms for a certain number of years. This will further enable the experience graduate to transfer skills to the newly graduates.

It is further recommended that there must be a comprehensive awareness and education to all South Africans. This could be done in different coordinated forms such as annual conference targeting school learners, University and college students and the community at large. The awareness could further be conducted at a local municipal area by the responsible local authority as a road show. The awareness could further be promoted through the community newspapers. This will enable the local residents to read and understand the significance, benefits and opportunities offered by the solar PV and CSP.

Research and Development

It is recommended that government should invest in the research and development through different coordinated ways. The research institutions such as SANEDI and CSIR should allocate sufficient funding to introduce variety and comprehensive solar PV and CSP technologies. It is believed that these institutions are unable to embark on comprehensive solar energy programme due to lack of funding. The government and private organisations should continue offering bursaries and scholarships to empower graduates to further their studies.

It is recommended that the government should clarify its renewable energy policy position, targets and objectives. The policy should further disaggregate the allocation of solar PV and CSP as to how it will be achieved.

References

Banks, D. and Schaffler, J. (2005), 'The potential contribution of renewable energy in South Africa', Johannesburg: Sustainable Energy and Climate Change Project (SECCP).

Chinese Renewable Energy Status Report (2009), 'Background paper: Chinese Renewable Status Report', Ren21, Available at: http://www.ren21.net/news38.asap.

Clean Energy Action (2007), 'Concentrated linear fresnel reflectors: the next generation of concentrating solar power', *Clean Energy Action fact sheet*, Available at: http://www.trecuk.org.uk/images/CLFR.pdf.

DME (2003), *White Paper on the Renewable Energy of the Republic of South Africa*, Department of Minerals and Energy, Pretoria, Available at: http://www.dme.gov.za/pdfs/energy/cabeere/cabeere_page4_2004.Pdf.

DME (2006), 'Mini-grid hybrid viability and replication potential', *Unpublished paper*, Department of Minerals and Energy, Pretoria, South Africa.

DME, Eskom and CSIR (2001), 'South African Renewable Energy Resource Database', Available at: http://www.csir.co.za/environmentek/sarerd/contact.html.

DoE (2011), 'Integrated Resource Plan for Electricity 2010-2030', *Unpublished paper*, Department of Energy, Pretoria, South Africa.

Digest of South African Energy Statistics (2009), Department of Energy, South Africa, Pretoria.

Edkins, M., Winkler, H. and Marquard, A. (2009), 'Large-scale rollout of Concentrated Solar Power in South Africa', Energy Research Centre, University of Cape Town, South Africa.

Environment America (2008), 'On the rise: solar thermal power and the fight against global warming', In: Del Chiaro, B., Payne, S. and Dutzik, T. (eds.), Environment America Research and Policy Center, Spring 2008, Available at: http://energy.sourceguides.com/business/byGeo/byC/SouthAfrica.

IEA, International Energy Agency (2008), *Energy Technology Perspectives*, Paris: IEA.

Mills, D.R. and Morrison, G.L. (1997), 'Advanced Fresnel Reflector Powerplants: Performance and Generating Costs',

Proceedings of Solar 1997, Australian and New Zealand Solar Energy Society.

New Energy Finance (2009), *Global Trends in Sustainable Energy Investment*, UNEP, UK (Endorsed by REN21).

REN21 (2009), *Renewables Global Status Report 2009 Update*, Paris: REN21 Secretariat.

REN21 (2010), *Renewables Global Status Report 2010 Update*, Paris: REN21 Secretariat.

Spellmann (2009), *The CSP industry: an awakening giant*, London: Deutsche Bank, Global Market Research, Companies.

Staley, B.C., Goodward, J., Rigdon, C. and MacBride, A. (2009), 'Juice from concentrate: reducing emissions with concentrating solar thermal power', World Resource Institute, Centre for Environmental Markets. Available at: http://pdf.wri.org/juice_from_concentrate.pdf.

Winkler (ed.) (2007), *Long Term Mitigation Scenarios: Technical Report*, Prepared by the Energy Research Centre for Department of Environment Affairs and Tourism, Pretoria, October 2007. Available at: http://www.erc.uct.ac.za/Research/publications/07-Winkler-LTMSTechnical% 20Report.pdf.

Chapter 10

The Current Status of Solar Energy Policy and Technology Development in Nigeria

Chigbogu G. Ozoegwu[1]

1. Introduction

The renewable energy (RE) resources which include solar, biomass, wind, geothermal and hydropower are environmentally friendly, readily/universally available and infinite in extent or replenished (or can be replenished) with usage, unlike the conventional energy resources which lack these benefits. As highlighted in REN21 (2014), other benefits of RE include reducing health and environmental impacts of fossil energy use, boosting energy access and security as well as providing secondary benefits such as improved opportunities for education, job creation, rural economic development, poverty reduction, and gender equality. Motivated by these benefits, policies and laws have been formulated to drive RE development and, as a result, RE has been rising in importance in the world energy mix. RE constituted 13% of the world's

1 Department of Mechanical Engineering, University of Nigeria, Nsukka, Nigeria. Email: chigbogug@yahoo.com & chigbogu.ozoegwu@unn.edu.ng

primary energy demand in 2011 (The International Energy Agency, 2013) and 19.3% of global final energy consumption as of 2015 (REN21, 2017). By the end of 2016, the estimated renewable energy share of global electricity production was 24.5% (REN21, 2017). RE accounted for 62% of net additions to global power generating capacity, attaining almost 2,017 GW, in 2016. The most important benefit is the limitless availability which can be appreciated against the very limited availability of the major fossil energy resources presented in Table 1. In the table, t_L is the length of time the reserves would last if production were to continue at the 2013 rate (British Petroleum, 2014). In 2016, about 9.8 million people were employed by the renewable energy sector, and 31.5% of them were employed in the solar PV sub-sector (REN21, 2017). These few facts illustrate the prospective global outlook of RE.

Table 1: The global proven reserves of major fossil fuels in 2013

Fossil fuel type	World reserve	[years]
Oil		53.3
Natural gas		55.1
Coal		113

Solar energy is the top RE resource because it is the source of the other RE resources. Furthermore, its universal availability, almost-zero environmental impact on utilization, risk-free supply facilities and the average daily trend that somewhat follows the trend of human electricity demand make solar the supreme RE resource (Bradford, 2006). For this reason, solar energy has greater competitiveness in the long-term RE targets of countries (Ozoegwu, Mgbemene, & Ozor, 2017). The intense R&D together with the evolving policy/legal landscape, and greater competitiveness have improved cost of manufacture and operational efficiency of solar devices. This resulted in reducing the price of PV panels, elongating the service life, reducing the space requirements of PV panels per energy supply and, ultimately, enhancing the global solar energy market. The price has fallen from US$ 30/W about 30 years ago to below US$ 1/W in 2013 and the price continues to fall as average module prices fell by about 29% to US$ 0.41/W between the fourth quarters of 2015 and 2016 (REN21, 2017). Most panels available in today's market have efficiencies in the neighbourhood of 15% while triple-junction (GaInP/GaAs/Ge [germanium]) cells and concentrating PV have been reported to attain efficiencies of about 44%. As a result, the current global market for solar PV technologies is increasing at a rate of more than 35% yearly. Solar PV accounted for about 47%, (almost half!) of newly installed renewable power capacity in 2016 followed by wind which contributed 34% (REN21, 2017). These points illustrate that solar energy is key to sustainable development and, thus, it must continue to enjoy favourable fiscal policies and laws. Reduction in the cost of solar energy is still sensitive and vulnerable to policy changes and measures to protect fossil fuels by exporting countries. For example, it will not be surprising for Nigeria to seem to delay passing laws to implement solar energy policies that may compromise crude oil market, thus there is a need to highlight the long term benefits solar energy offers. This is because Nigeria is a mono-economy almost solely dependent on petroleum.

2. A Nigerian Outlook

Nigeria, with a landmass of 923,768 km2, is the seventh most populous nation in the world after China, India, USA, Indonesia, Brazil and Pakistan. The population in the year 2006 was 140,431,790 and the population growth rate is at 3.2% per annum (National Population Commission (NPC) [Nigeria] and ICF International, 2014), meaning that a rational estimated population for the year 2020 is 218,263,539. This is an enormous, but yet untapped, market for modern conventional and alternative energy technologies. Eighty percent of total primary energy consumption in Nigeria of about Btu (EJ) in the year 2014 was from open (very inefficient) burning of traditional biomass (e.g. woodfuels and agricultural byproducts). Therefore, building a developed Nigeria that is driven by RE (especially solar) is a veritable way to reduce greenhouse gas emission and create a massive trade hub for emission credits. Fortunately, Nigeria is heavily endowed in both fossil and RE resources as summarized in Tables 2 and 3. By the end of 2016, Nigeria's crude oil reserve was 37.5109 barrels (British Petroleum, 2018). Yet, Nigeria is embroiled in crises of final energy. The share of Nigeria's people with access to electricity in 2014 was 45% with a target of 75% set for 2020 and 90% set for 2030 (REN21, 2017). The share has grown to 61% in 2016 (REN21, 2018), indicating an improvement. The crises in the Niger-Delta region of Nigeria is a battle for control of Nigeria's economic mainstay; the oil and gas sector. It is a battle that seems like a must-win for Nigeria's government since the sector typically accounts for 75% and 95% of total government revenue and total export revenue respectively (Ozoegwu et al., 2017). Moreover, Nigeria is desperate to protect gas pipelines supplying the few operational electricity generators from being damaged by the militants. This is because amongst the existing 23 power generating stations in Nigeria, 20 are gas-powered while the remaining three are hydro-powered (GOPA-International Energy Consultants GmbH, 2015). It will be difficult to think of Nigeria meeting the full electrification rate through grid expansion. Many terrains in Nigeria, like coastal areas and creeks, are too difficult or too isolated to be covered by grid power, leaving only the distributed power option. Moreover, it is not feasible to continue expanding Nigerian grid because of insufficient national power generation capacity that frequently fluctuates from a relatively high of around 3,700 megawatts to lower values and, sometimes, to total failure. This is against the national dream to increase generation capacity to more than 25,690 MW by 2020, mainly from fossil-fuel sources and hydro (U.S. Energy Information Administration (EIA), 2016). Renewable energy, especially solar, is therefore the only option for possible 100% electrification in Nigeria. Fortunately, Nigeria is geographically favoured for solar energy economy. Nigeria has a landmass of 923,768 km2 that receives daily solar irradiation of 3.5 kWh/m2 (145.83Wm-2) in the south and 7.5 kWh/m2 (312.50Wm-2) in the north (Aliyu, Dada, & Adam, 2015). The solar energy resource is much above the global average ground-level solar irradiance of 170 Wm-2(World Energy Council, 2013). The importance of solar energy for sustainable development of Nigeria has been identified and, as a result, many researchers have tried to assess solar energy availability in Nigeria (Fadare, 2009; Ozoegwu, 2018; Ozoegwu, 2018; Swartman & Ogunlade, 1967). Ozoegwu (2018) reviewed most of these works that assessed solar energy in Nigeria.

Table 2: Nigeria's fossil fuel energy reserves

Resource type	**Reserves**	
	Natural units	**Energy units** (Btoe)
Natural gas	187 trillion SCF	4.19
Crude oil	36.22 billion barrels	5.03
Tar sand	31 billion barrels equivalent	4.31
Coal and lignite	2.175 billion tonnes	1.52

Source: Ozoegwu et al. (2017)

Table 3: The renewable energy potential of Nigeria

Resource type		Reserves	Units
Small hydropower		3500	MW
Large hydropower		11,250	MW
		11,235	
Wind		2–4	m/s at 10m height (mainland)
Solar radiation		3.5–7.0	kwh/m²/day
		3.5–7.5	
Municipal Solid wastes	Generated	17451000	tonnes per year
	Collected	7329000	
Animal waste		211 million assorted animals (285.065 million tons/yr of production)	
		61	million tons/yr
Energy crops and agric residue		28.2 (30% of total land)	million hectares of arable land
Crop residue		83	million tons/yr
Fuel wood		11	million hectares of forest and wood land
Wave and tidal energy		150,000	TJ
		1759.6	toe/yr

Source: Ozoegwu et al. (2017)

3. The current status of solar energy technology development in Nigeria

Grid-connected Solar Photovoltaic

Solar radiation can induce forward-biased current at the p-n junction of semiconductors to effect the direct conversion of solar radiation to electricity. This is because, on the absorption of photons of solar radiation, transition of electrons across the narrow-forbidden band to the conduction band occurs to create the junction current which can be used directly or saved in batteries or fuel cells. This is the basis of operation of solar photovoltaic (PV). Solar photovoltaic is gaining ground in Nigeria. In 2015, the Nigerian Energy Support Programme surveyed 53 companies and found that they have installed 115 MW of off-grid photovoltaic which includes mini-grids (backup to grid supply) and stand-alone systems (GOPA-International Energy Consultants GmbH, 2015). Solar PV is popular for street lighting. A hybrid mini-grid of 10 kW solar PV and 2 kW wind power has been jointly installed by the Sokoto Energy Research Centre, the World Bank, the Energy Commission of Nigeria and the Sokoto Government in Danjawa Village, Sokoto. A hybrid plant of 10 kW solar PV and 15 kW wind has been inaugurated at the NAPTIN Regional Training Centre in Kainji, thanks to the funding by the German Federal Foreign Office and implementation by the Deutsche Gesellschaft für Internationale Zusammenarbeit (GIZ). The reason for the growing acceptance is the acute shortage of grid power in Nigeria and massive dependence on petrol/diesel generating set which has been proven to be of higher levelized cost than solar PV. As far back as in the 1990s, a comparative economic analysis of three alternative sources of electricity (diesel generator set, solar PV, and national grid connection) to remote rural areas in Nigeria favoured solar PV (Bugaje, 1999). This point should be appreciated against the favourable backdrop that the cost of solar PV electricity has fallen considerably since then to about USD0.41/W (REN21, 2017). Among the courses of study and training being developed by GIZ is solar PV installation. The current level of PV installation in Nigeria (with about 53% of population of ECOWAS bloc), is far below when compared to other countries of the bloc such as Cape Verde (with just 0.2% of ECOWAS population) that has installed 7.5 MW of grid-connected solar PV plants (ECOWAS Centre for Renewable Energy and Energy Efficiency, 2013) and Ghana (about 7.7%

of ECOWAS population) which has set up the Navrongo PV plant in 2013 with 1.92 MW capacity (REN21, 2014). There are prospects in the Nigerian power sector which could be traced to the Power Sector Roadmap introduced by the last administration of the federal government, which successfully privatized the sector and awarded licenses to more efficient profit-driven players in the sector. Nigerian Electricity Regulatory Commission (NERC) issued licenses to a number of investors as of 2014 for development of various solar PV power projects. The companies together with the capacities and locations of the projects are summarized in Table 4. Other solar PV power projects which are planned or under development in Nigeria have been discussed in Ozoegwu et al. (2017) and summarized below. In September 2011, a 50 MW solar farm was launched by the Synergent Powersh Group of Company in Kaduna state. Because the hydropower generation capacity of Shiroro Hydroelectric Power Station in the Niger State of Nigeria traditionally declines during dry seasons, the management has proposed a back-up PV solar installation of 300 MW for off-peak seasons. The New Horizon Energy Resources announced a plan of building a 100MW solar plant in Nasarawa. In 2014, SkyPower FAS Energy signed a Memorandum of Understanding (MoU) with the Federal Government of Nigeria and the Delta State Government to construct a 3000 MW of utility-scale solar photovoltaic (PV) project over the next five years at different Nigerian locations which include the Delta State. Spurred by environmental pollution caused by crude oil production in the Delta state, the state government signed a MoU with Yutal Li Nigeria Limited to construct of a 100MW solar power plant. Nigeria has also signed a MoU with Super Solar for the construction of plants to produce 3000 MW of solar power.

Table 4: The name of the companies, capacity and location of the solar power projects backed by NERC licenses.

Licensee	Capacity (MW)	State	Geopolitical Zone
Rook Solar Investment Limited	50	Osun	South-West
Quaint Global Nigeria Limited	50	Kaduna	North-West
Nigeria Solar Capital Partners	100	Bauchi	North-East
Anjeed Kafanchan Solar Limited	10	Kaduna	North-West
Lloyd and Baxter LP	50	Abuja	North-Central
KVK Power PVT Limited	50	Sokoto	North-West
Pan African Solar	54	Katsina	North-West

Source: GOPA-International Energy Consultants GmbH (2014)

In essence, Nigeria has a near-zero share in the impressive global grid-connected solar PV power generation. The estimated global grid-connected PV power capacity by the end of 2010 was 34GW while SolarPower Europe has indicated that the global cumulative capacity of PV installation reached 178 GW in the year 2014 (Ozoegwu et al., 2017). The trend from 1995 to 2014 of global installed PV capacity which was graphically analyzed and discussed in Ozoegwu et al. (2017) is updated in what follows. Figure 1 shows an enormous growth of cumulative global installed PV capacity from 3.7GW in 2004 to 402GW in 2017. The cumulative capacity before 2004 was low. Figure 1 also shows the yearly additions to global installed capacity since 2004. It shows that yearly additions have continued to grow from year to year. The percentage year to year growth rate of cumulative global installed PV capacity is shown in Figure 1 as trend (a). A data for *i*th year on the trend is given as $r_{i,y2y}=100[(x_i/x_{i-1})-1]$. It is seen that a net positive growth occurred from 2004 to a very high value above 70% in 2011. Thereafter, there has been a sustained decrease in the year-to-year growth rate from 75% in 2011 to 28.26% in 2014, followed by a somewhat steady rate of about 33%. The average growth from any previous year to 2017 is shown in Figure 1 (trend b). This trend is used to highlight the constant yearly growth rate from any previous year of interest to attain the cumulative global installed capacity of the year 2017. Since constant year on year growth is assumed in trend (b) the data for *i*th year is given as $r_{i,py2cy}=100[(x_{2017}/x^{i})^{1/(2017-i)} -1]$. It means that relatively low and high yearly addition will cause a rise and fall respectively in r_{py2cy}. The trend fluctuated slightly above 40% before 2010 and has fluctuated slightly below 40% after 2010. Figure 2 shows the trends of growth of global installed PV capacity between 2002 and 2017. This is also shown by Figure 1 (trend c). It is evident that globally installed PV capacity grew significantly since 2013. Nigeria must begin to implement supportive policies and start transforming plans in order to have a noticeable share in the global solar PV power and meet its year 2025 large-scale (>1MW) solar PV power target of 500MW (REN21, 2018).

Figure 1: The trend of global (grid and off-grid) installed PV capacity

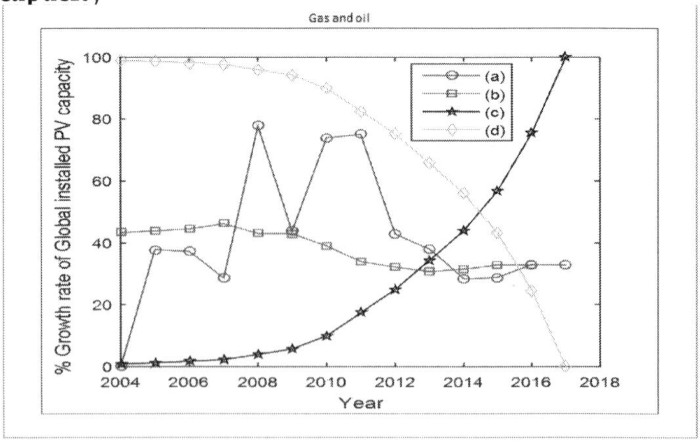

Source: REN21, 2018

Figure 2: The trends of growth of global installed PV capacity

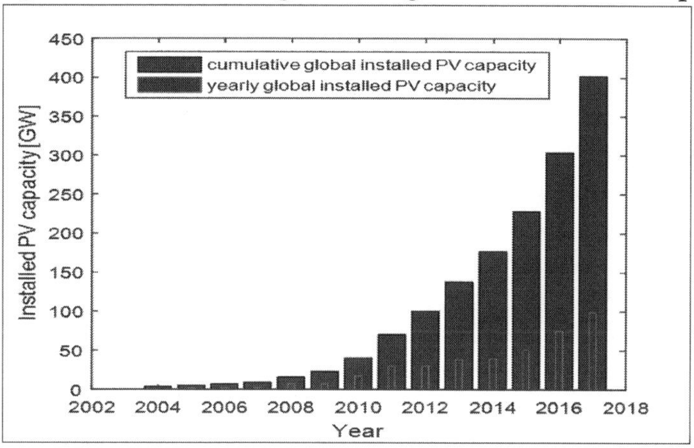

Source: REN21, 2018
Note:
(a) Percent year to year growth rate,
(b) average percent growth rate from Year to 2017
(c) Percentage of installed capacity by end of Year to installed capacity by end of 2017 and
(d) Percentage of installed capacity since the beginning of Year to the end of 2017.

Grid-connected Solar thermal

Bradford (2006) defined solar thermal applications as all those that use solar radiation to power heat-driven mechanisms such as process or conductive fluids, solar cooking, agricultural drying, desalination, and cooling. Nigeria has an interest in active solar thermal for power generation. For power generation, high-pressure steam is usually superheated by concentrating solar radiation on an absorber. Three types of concentrating collectors are known (Tian & Zhao, 2013); heliostatfield collector, parabolic dish collectors and parabolic trough collectors. Heliostat field collector consists of an array of flat mirrors or heliostats directing incident solar radiation to an absorber. Increasingly, solar thermal power plants are being deployed for the supply of electricity to the grid. Solar thermal power utilizes the established technologies of heat engine cycles which can run on fossil fuels as back-up during low or zero solar energy availability. Solar thermal power is less affected by the intermittency of solar radiation because of the feasibility of high capacity thermal storage. As a result of the above relative benefits of solar thermal plants, they have a lower net and levelized cost compared to grid-connected solar PV generation.

Nigeria has no grid-connected solar thermal power generation and shows no evidence of prospects (Ozoegwu et al., 2017). SolarPACES maintains a database on concentrating solar power (CSP) projects that are either operational, under construction or development around the world (National Renewable Energy Laboratory, 2018). The database does not mention Nigeria thus indicating that Nigeria's zero status and dismal prospects in grid-connected solar thermal power remains. Only four African countries (Algeria, Egypt, Morocco and South Africa) are featured in the database. Based on the information extracted from the database, the current operational solar thermal power projects in Africa are summarized in Table 5, while those under construction or development are summarized in Table 6. The newest operational concentrated solar thermal power projects in Africa are NOOR1 and NOOR 2 in Morocco which have a combined capacity of 360 MW and Khi Solar One, Bokpoort and Kathu Solar Park in South Africa with a combined capacity 205 MW.

The trend of the cumulative and yearly global installed Concentrated Solar Thermal Power (CSTP) capacity from 1984 to 2017 is given in Figure 3. It is seen that on average the cumulative global installed capacity grew from year to year since 1984 and it is quite discouraging that Nigeria, the most populous and biggest crude oil producer in Africa, is not part of this success story. The trend for yearly global addition of installed capacity is also included in the figure to show that, except for the stagnation period, yearly addition which mainly grew from year to year since 1984 seems to be falling in recent times.

Table 5: The currently operational solar thermal power projects in Africa

Name	Capacity (MW)	Country	Location	Technology Type	Thermal storage
Kaxu Solar One	100	South Africa	Pofadder	parabolic trough	molten salts
	Started operation in 2015; 775 acres of Solar field; supplies 80,000 Households; Eliminates 300,000 Tons of CO2 emissions per year. Over 1,000 jobs created at peak during construction. Currently employs 80 people in full-time, long-term, high-paying O&M jobs				
ISCC HassiR'mel	20	Algeria	HassiR'Mel	parabolic trough	
	Started operation in 2011; 1.94 million sq. ft of Solar field; Thermal oil as heat transfer fluid				
ISCC Ain Beni Mathar	20	Morocco	Ain Beni Mathar	ISCC with parabolic trough	
	Started operation in 2010				
ISCC Kuraymat	20	Egypt	Kuraymat	ISCC with parabolic trough	
	Started operation in 2011; saving 5,000 tonnes of natural gas per year (180,000 tonnes CO2/year); Therminol VP-1 as heat transfer fluid				
Airlight Energy Ait-Baha Pilot Plant	3	Morocco	Ait Baha	parabolic trough	with 5h heat storage
	Started operation in 2014; Packed-bed of rocks as Thermal Storage material				
NOOR I	160	Morocco	Ouarzazate	parabolic trough	3h heat storage using Molten Salt
	Started production in December 2015; Dowtherm A as heat transfer fluid; A commercial power station				

NOOR 2	200	Mo-rocco	Ouarzaza-te	parabolic trough	7h heat storage using Molten Salt
	Started production in January 10, 2018; Thermal oil as heat transfer fluid				
Khi Solar One	50	South Africa	Upington	power tower Superheated steam	2h heat storage using molten salts
	Started production in February 5, 2016; Water/Steam as heat transfer fluid; A commercial power station				
Bokpoort	55	South Africa	Grobler-shoop	parabolic trough	9.3h heat storage using molten salts
	Started production in March 14, 2016; Dowtherm A as heat transfer fluid; A commercial power station				
Kathu Solar Park	100	South Africa	Kathu	Parabolic trough	4.2h heat storage using molten salts
	Started production in February 23, 2018; Thermal oil as Heat-Transfer Fluid; A commercial plant				

Source: National Renewable Energy Laboratory (2018)

Table 6: The solar thermal power projects under construction or under development in Africa

Name	Capacity (MW)	Country	Location	Technology Type	Thermal storage	Notes	Status date
eCare Solar Thermal Project	1	Morocco	-	Linear Fresnel reflector	2h heat storage using Steam drum	Expected to be completed in 2014; Water as the Heat-transfer fluid	November 27, 2012
NOOR 3	150	Morocco	Ouarzazate	Power tower	8h heat storage using Molten Salt	Under construction. Expected to be completed in 2017; Molten salt as Heat-Transfer Fluid	March 23, 2017
IRESEN 1 MWe CSP-ORC pilot project	1	Morocco	Benguerir	Linear Fresnel reflector	20 minutes buffer thermal storage	A research project. Under construction. Expected to be completed in September 2016; Mineral oil as Heat-Transfer Fluid	July 26, 2016
Redstone Solar Thermal Power Plant	100	South Africa	Postmasburg	power tower	with 12h heat storage using molten salts	A Commercial Project under development. Expected to be completed in 2018; Molten salt as Heat-Transfer Fluid	September 8, 2016

Xina Solar One	100	South Africa	Northern Cape	parabolic trough	5h heat storage using molten salts	Under construction; Expected to be completed in 2017; 90,000 Households to be supplied; 398000 Tonnes of CO_2 emissions to be eliminated per year	April 14, 2015
Ilanga 1	100	South Africa	Upington	Parabolic trough	with 4.5h heat storage using molten salts	A project under development; Expected to be completed in 2020; Thermal oil as the Heat-transfer fluid	May 22, 2017

Source: National Renewable Energy Laboratory (2018)

Figure 3. The trend of global installed CSTP capacity

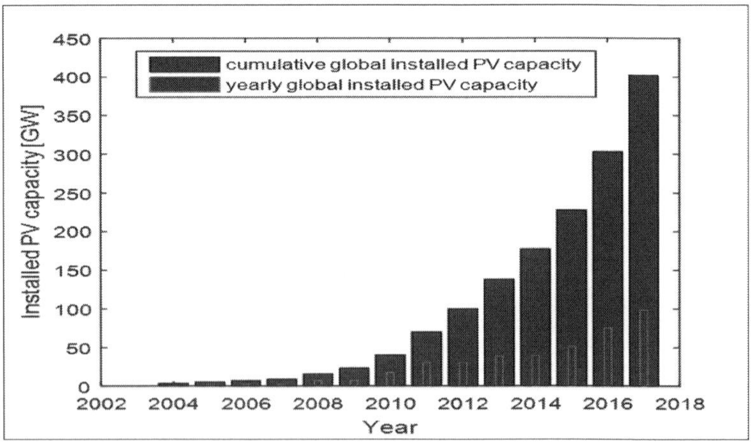

Source: REN21, 2018

The percentage year-to-year growth rate of cumulative global installed CSTP capacity is shown in Figure 4 (trend a). It is seen that the years 2010, 2011 and 2012 witnessed rising growths of 38.02%, 46.30% and 61.39% respectively while 2012 to 2017 witnessed a declining growth. The trend suggests a relatively high yearly addition up to 1988, a sustained relatively low yearly addition from 1988 to 2008 (the yearly additions is zero for most of the years) and then, a relatively high yearly addition to 2014. The yearly addition has been falling since 2008 to the present. Average percentage growth from any past year to attain the installed capacity by the end of 2017 is shown as trend (b) in Figure 4. This trend is useful to highlight the constant yearly growth rate to be maintained from any past year of interest to attain the cumulative global installed CSTP capacity of year 2017. For example, the constant yearly growth rate of 19.43%, 12.11%, 22.42%, 24.15%, 3.72% and 2.08% were respectively needed from the end of 1984, 1994, 2004, 2010, 2014 and 2016 to meet the cumulative capacity by the end of 2017. Trend (c) shows the percentage of installed capacity by end of a past year to installed capacity by end of 2017. It can be seen that 0.28%, 7.21%, 7.21%, 9.88%, 22.00%, 89.61% and 97.96% of all installed CSTP global capacity by the end of 2017 were installed by the end of 1984, 1994, 2004, 2008, 2010, 2014 and 2016 respectively. It is interesting to state that 90.12% of all globally installed CSTP came on board within the last decade. This highlights the monumental market growth rate. It can be seen from trend (d) that, by the end of 2017, about 99.7%, 92.8%, 92.8%, 78%, 10.4% and 2% of all globally installed CSTP capacity came on board since the beginning of 1985, 1995, 2005, 2011, 2015 and 2017, respectively. Nigeria must begin to implement supportive policies, promote technical know-how and start implementing plans in order to have a share in the global concentrated solar thermal power and meet its year 2025 concentrated solar power target of 5 MW (REN21, 2018).

Figure 4. The trends of growth of global installed CSTP capacity.

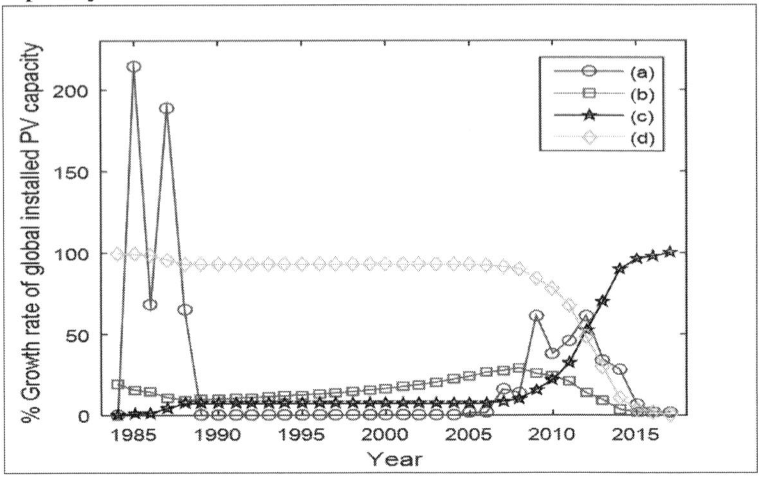

Source: REN21, 2018
Notes:
(a) % year to year growth rate
(b) average % growth rate from Year to 2017
(c) % of installed capacity by end of Year to installed capacity by end of 2017
(d) % of installed capacity since the beginning of Year to the end of 2017

Status of off-grid solar integration

Solar energy is mainly used non-technologically (open air) for drying purposes. Open air solar drying is applied to items such as clothing, furniture, agricultural food crops, agricultural crop residues, forest biomass and products of livestock husbandry. Technological applications in Nigeria are demonstration projects. They are low-energy capacity solar street lighting, solar refrigerators, solar-powered water pumps and stand-alone power supply. Such low-capacity solar installations or projects in Nigeria have no comprehensive database. As reviewed by Ozoegwu et al. (2017), the projects include:

- Solar-powered lighting of more than 100 streets in the capital of Nigeria, Abuja.

- Fifty-eight solar projects (out of which 50 are tagged "Solar Based Rural Electrification") undertaken by the Energy Commission of Nigeria (ECN). The status of the projects (planned, still under construction or completed) were not indicated.
- Solar electrification for twenty-two communities with a combined population of over 70,000 in remote areas in the Zamfara State of Nigeria.
- Six hundred projects installed by the Rural Electrification Agency (REA). Information regarding the types of technology, capacity and consumers supplied was not given.
- Various non-estimated off-grid solar energy penetration under the NAIJA LIGHT Solar Electrification Programme which include: (i) the NAIJA LIGHT Solar Kits (handy and portable solar 50W and 100W kits equipped with a light bulb, an outlet for a fan and 10 point mobile phone charging unit); (ii) NAIJA LIGHT Solar Hair-care Kits (300MWh kits equipped with two lighting points and a 10point mobile phone charging unit each; and (iii) NAIJA LIGHT Primary Health Care Solar Kits (mini freezer for storage of serums and vaccines, two lighting points and a 10 point mobile phone charging unit for rural level healthcare delivery).

Made-in-Asia solar phone chargers and solar-powered rechargeable lanterns are other popular solar devices in Nigerian markets. Across Nigeria, solar PVs are increasingly being mounted on rooftops and specialized frames as stand-alone systems or connected as hybrid systems with the National grid. The hybrid systems are intended to reduce or pre-empt grid-electric power charges by being equipped with batteries that are charged by solar PVs during solar availability and charged by grid power during solar unavailability.

The most recent channels of penetration of solar technology in Nigeria include (REN21, 2018): (i) a Nigerian company producing solar-powered tricycles; (ii) A business partnership (by Renewvia Energy, Community Energy Social Enterprises Limited and microfinance company Kilowatts) providing power to 10,000 Nigerian homes with solar mini-grids using a pay-as-you go scheme in 2017; and (iii) World Bank initiative to construct mini-grids in several African countries including Nigeria.

To sum up, Nigeria appears to have seen significant penetration of the stand-alone solar energy technology systems.

Indigenous Capacity in Solar Manufacture Technology

Efforts on indigenous manufacturing of reliable PV modules have been underway. A facility partly financed by the World Bank was completed in February 2014 in Sokoto by a German company to produce solar panels designed to operate in extreme climates. Further, under the intervention programme of the National Agency for Science and Engineering Infrastructure (NASENI), the Karshi Solar Panel Plant (KSPP) was built to annually manufacture 7.5 MWp. Though NASENI boasts of producing about 2,800 pieces of solar panels since 2011 (Giwa, Alabi, Yusuf, & Olukan, 2017), the impact of these efforts has not been felt as no sufficient national manufacturing facilities have been achieved in Nigeria, only some local production of PV modules is done currently (GOPA-International Energy Consultants GmbH, 2015). Solar PV modules are subjected to location-specific thermal-mechanical excitations which induce failure of corrosion and fatigue/creep types in the system of weak linkages holding the cells together. As a result, solar PVs designed and manufactured on the basis of European and Asian standards are known to fail much earlier than expected when installed in Nigeria. This highlights the need for indigenized design, standardization and manufacture of solar PVs intended for applications in Nigeria. Therefore, the Nigerian academia must build capacity in computational simulation of the thermal-mechanical processes/effects induced by Nigerian environment on solar PV modules in order to establish thermal, material/manufacturing, geometric and accelerated lifetime testing standards for domestic production of solar PV suited to the local environment. In summary, though NASENI and Sokoto Energy Research Centre (SERC) have started production of solar panels in Nigeria, a lot of work still needs to be done in the development of solar technology equipment, standards for material, design and equipment manufacture.

The Nigerian research community has excellent knowledge of the working principles of solar thermal technologies. There is no evidence of local manufacture of concentrated solar thermal in Nigeria. Non-concentrated solar thermal technologies like domestic cooking, heating, cooling and drying are designed and constructed all over Nigeria by researchers mostly for experimental purposes. Some of the laboratory-sized constructions involving personal participation are shown in Figure 5. From an economic, technical, environmental and health points of view, non-concentrated solar thermal technologies, especially cooking, heating and drying are likely to be installed and used widely than the solar PV or concentrated solar thermal technology. The economic and environmental advantages can be viewed in terms of very low capital investment and very low life-cycle and post-life cost. This is because the end-user products of such applications, unlike the solar PV, do not constitute much environmental and disposal problems. Though the issue of variability of solar energy plagues these applications, free solar availability together with the ultra-low maintenance/environment cost could help solar thermal out-compete the conventional options in the short-term. It was stated earlier that 80% of the total primary energy consumed in Nigeria in the year 2014 was from open burning of traditional biomass such as wood, agricultural bi-products, and animal wastes. In 2015, the share of Nigerian population without access to clean cooking was 94% (171 million people) and, unfortunately, Nigeria has not set a target for the share of its population with access to clean cooking (REN21, 2018). The energy consumption is mainly for heating and cooking; thus, promoting the simple solar thermal technologies could help reduce the negative health implications of the massive dependence on open burning of traditional and waste biomass. Nigeria must set the target of

100% access to clean cooking for its population by 2030, which is the target set by several other African countries like Ghana, Angola, Swaziland, and Rwanda.

Figure 5: Laboratory-sized construction non-concentrated solar thermal devices

(a) Solar water heating collector

(b) Solar box cooker

(c) Solar drier

4. The solar energy policy development in Nigeria

The global RE policy landscape has penetrated 154 countries by 2015, compared to 144 in 2014 (REN21, 2015). Countries have focused much more on RE Regulatory Policies for the power sector than the transport as well as heating and cooling sectors. As of 2017, 128 countries have enacted Regulatory Policies, which is an increase from 116 countries in 2015 (REN21, 2018). For the transport sector, the number of countries with regulatory policies went up from 65 in 2015 to 70 in 2017. The number of countries with regulatory policies in heating and cooling remains constant at 29 since 2014.

Nigeria has master plans, action plans and policies for the promotion of RE. To stimulate an enhanced penetration of RE in Nigeria's energy mix, economic and financial incentives were proffered to specifically reduce the initial high cost of investment in RE technologies in the November 2012-revised edition of Nigerian Renewable Energy Masterplan (REMP). From 2013 to 2014, the Federal Ministry of Power drafted the National Renewable Energy and Energy Efficiency Policy (NREEEP) which was approved by the Federal Executive Council in May 2015. Under NREEEP, solar electricity targets were set at 117, 1343 and 6831MW by 2015, 2020 and 2030 respectively (Federal Ministry of Power, 2015). The NREEEP created a strategic framework for relevant Ministries, Departments and Federal Agencies to develop relevant policies, regulations and economic instruments. These included Mandatory or Voluntary Renewable Portfolio Standards (RPS), Generation Disclosure Requirement (GDR), Power Production Tax Credit (PTC), Feed-in tariffs (FIT), Adoption of a Public Benefits Fund (PBF), Capital Grants, Tax Holidays, exemptions and other incentives for RE projects. Renewable Energy support policies which are in place in Nigeria are (1) Feed-in tariff /premium payment, (2) Capital subsidy, grant, or rebate, (3) Reductions in sales, energy, CO2, VAT, or other taxes, (4) Public investment, loans, or grants, (5) Biofuels obligation/mandate (REN21, 2015). In addition, tendering, electric utility, quota obligation, Renewable Portfolio Standards (RPS), and Renewable Energy in Intended Nationally Determined Contribution (INDC) are some support policies that have been introduced in Nigeria (REN21, 2018). These RE supportive policies have been discussed in detail by Ozoegwu et al. (2017) as being applicable to solar energy. Hereunder, the discussion is updated with the latest developments.

Feed-in tariffs are premium payments that guarantee reasonable returns from investments in new and RE power technologies. FITs usually place higher premium payments for costlier RE power technologies. For example, the FITs stipulated in the Multi Year Tariff Order (MYTO) II by the NERC charged ₦92192 and ₦33433 per MWh for Solar PV plants and land-mounted wind power plants for the year 2016. About eight months after the NREEEP was created (precisely on the 8th of December 2015), the Feed-in-Tariff Policies of NERC for Renewable Energy Sourced Electricity in Nigeria was approved by the government. The document stipulates FIT of US$ 177.00 / MWh or ₦ 35,399.54 / MWh for solar PV (1 to 5 MW) from the 2016 base year. There is a provision for review of the FIT every three years, only for new projects. The FIT is much more attractive to the Nigerian Bulk Electricity Trading Plc (NBET) than the earlier, much costlier stipulation of ₦92,192 in MYTO II, which NBET rejected (GOPA-International Energy Consultants GmbH, 2015). The NBET is the transitional public liability company that was instituted to buy bulk electricity from the GENCOs and sell to the DISCOs. The new document stipulates a purchase obligation on the Off-Takers (like the DISCOs) in favour of RE sourced power by requiring NBET or its successors to purchase 50% of the system-established renewable electricity capacity and the Off-Takers to purchase the remaining 50%. The inability of an Off-Taker to cope with its obligation is to be met with a financial penalty. This extant regulation means that there is hope for solar PV based grid expansion in Nigeria. The problems of FITs stipulated by NERC in MYTO II are high cost, lack of priority access to electricity from renewable energies and lack of obligations on the Off-Takers. The major problem which could discourage foreign investors is the uncertainty created by the clause that NERC may amend or repeal, in whole or in part, the provisions of the regulation. The government has approved comprehensive guidelines which provide clarity about key regulatory aspects important to developers, such as tariffs and grid integration, to regulate the deployment of mini-grids (REN21, 2018). This existing FIT could make such mini-grids more attractive, as a system is already in place in case they may want to sell to the grid.

Nigerian policies on capital subsidy, grants, or rebates, have not witnessed appreciable changes beyond what was discussed in Ozoegwu et al. (2017). NREEEP provides soft loans and special low interest loans from the Power Sector Development Fund for renewable energy projects. As part of the strategies for RE Financing, the Nigerian Government is committed to provide grants to the local communities for renewable energy projects, from the proceeds from the privatization of PHCN infrastructure. The strategy by the government to attract international donors and NGO's is the funding of demonstration RE projects and feasibility studies through the provision of grants and donations. The major shortcoming of Nigerian policies on capital subsidy, grant, or rebate is the lack of obligatory quantitative figures. This shortcoming should be addressed, especially for solar power investments.

The discussion and recommendations on Nigerian public investment policy in Ozoegwu et al. (2017) are summarized here. The enactment of the Electric Power Sector Reform Act (EPSRA) in 2005 provided legal grounds for the unbundling of National Electric Power Authority (NEPA) and privatization of successor companies. As a result, the generation and distribution sectors of the Nigerian grid electricity sector have been fully privatized and the transmission sector, which has been under concession, is in the process of privatization. In NREEEP, the Government commits to the possibility of public-private partnership project models for accessing rural electrification funds, thus, preferring models that are based only on the private sector. Given how government monopoly has failed Nigeria in various sectors, the plans for full privatization of power sector is commendable, but there is still the need for public investment to surmount perceived high capital cost of the solar power sector (especially for rural electrification).

The NREEEP document provides a basis for the introduction of various Tax Incentives. These include tax holidays and exemptions for RE projects, power production tax credit to generators of renewable electricity, tax and duty exemptions for investors in the renewable energy sub-sector, tax deductible for result oriented RE research and development costs and exemption of income tax or duty for carbon credit trade. Tax incentives provided in NREEEP to promote the widespread use of RE include a five-year tax holiday for manufacturers of RE products to start from the date of commencement of manufacturing and five-year tax holiday on the dividend from investments on domestic RE sources. Exemption duty, in the form of free custom duty for two years, was set up in the document in favour of imported equipment and materials used in RE projects. Quantified commitments are lacking on the Tax Incentives and this has to be corrected especially in favour of solar. The above tax and duty incentives on RE products were also extended to energy efficient appliances and lighting in NREEEP (Ozoegwu et al., 2017). It has been recommended in Ozoegwu et al. (2017) that solar power investments to supply electricity to isolated localities be accorded "pioneer status" which is an extant law of the Federation of Nigeria contained in the Nigeria Investment Promotion Commission Act, Cap 117 No. 16 of 1995. This confers a seven-year tax holiday to industries located in economically disadvantaged local areas of Nigeria. Nigeria has approved tax rebates for manufacturers of solar panels, batteries, solar home systems, light emitting diodes and other components that support solar energy production under the new pioneer status law (OFFGRID NIGERIA, 2017)

Tendering is a way to procure RE capacity from competitive bidding of sellers who offer the lowest prices they may accept and the buyer reaching a contract with a seller based on price and non-price factors. Tenders, proposal projects and bids for solar energy and other REs in Nigeria are listed on the webpage Global Tenders (2018). The solar-related tenders include (i) the construction of 300 units of solar-powered motorized boreholes in communities, schools and health facilities; (ii) solar street lights from Kieprima to Mission Road in Opobo Town of OpoboNkoro LGA; (iii) supply of equipment and parts for installation and distribution of 30KW off grid solar rooftop electrification to University of Port Harcourt Teaching Hospital; and (iv) solar powered water scheme in Obuama Degema LGA. Most of the solar electrification tenders are summarized in Table 7. The Presidency of Nigeria approved Nigeria's INDC on 26 November 2015 for implementation from 2015 to 2030. Nigeria's INDC was formulated to contribute to the success of the Paris climate summit held in December 2015 in view of Nigeria's sustainable economic growth and reduction of carbon pollution (Federal Ministry of Environment, 2015). One of the main objectives of INDC is working towards achieving off-grid solar PV of 13GW.

Table 7: The solar electrification tenders in Nigeria

Project	Tender Notice Type	Action Deadline
Installation and distribution of 30KW off grid solar rooftop electrification to University of Port Harcourt Teaching Hospital	Tender notice	30 Nov 2018
Installation and distribution of 30KW off grid solar rooftop electrification to Bomadi General Health Hospital	Tender notice	30 Nov 2018
Installation and distribution of 30KW off grid solar rooftop electrification to Patani General Health Hospital Patani LGA	Tender notice	30 Nov 2018
Electrification/solar streetlights of Emjowhofee Avenue Along Western Delta Road in Ogharefe Ethiope West LGA	Tender notice	30 Nov 2018
Rural HT/ it electrification at Oda community in OzoroIsoo North LGA	Tender notice	30 Nov 2018
Installation and distribution of 30KW off grid solar rooftop electrification to Delta State University Teaching Hospital Oghara in Ethiope West LGA	Tender notice	30 Nov 2018
Installation and distribution of 30KW off grid solar rooftop electrification to Niger Delta University Teaching Hospital at Okolobiri, Yenagoa LGA	Tender notice	30 Nov 2018
Construction of solar mini grid and retrofitting of lights to energy efficient lights to power only lighting point in the House of Assembly in Delta State	Tender notice	30 Nov 2018
Construction of solar mini grid and retrofitting of lights to energy efficient lights to power only lighting point in the House of Assembly in Bayelsa State	Tender notice	30 Nov 2018

Construction of solar mini grid and retrofitting of lights to energy efficient lights to power only lighting point in the House of Assembly in Akwa Ibom State	Tender notice	30 Nov 2018
Construction of solar mini grid and retrofitting of lights to energy efficient lights to power only lighting point in the House of Assembly in Abia State	Tender notice	30 Nov 2018
Supply and installation of solar power generators in Health Center Facility in the 27 senatorial zones (lots 1-27)	Tender notice	30 Nov 2018
200MW solar power plant on Shendam Road	Project information	31 Dec 2019
0.5MW solar plant at Federal University of Petroleum, Delta State	Project information	31 Dec 2019
2MW solar plant at Nnamdi Azikiwe University, Awka	Project information	31 Dec 2019
1MW solar plant at Federal University Ndufu-Alike Ikwo in Ebonyi State	Project information	31 Dec 2019
3.5MW solar plant at Federal University of Agriculture Makurdi	Project information	31 Dec 2019
2MW solar hybrid plant at Usmanu Danfodiyo University Sokoto	Project information	31 Dec 2019
3MW solar hybrid plant at Bayero University Kano	Project information	31 Dec 2019
0.5MW solar hybrid plant at Abubakar Tafawa Balewa University	Project information	31 Dec 2019
Jigawa Solar Project	Project information	31 Dec 2019
50 MW solar plant at Kafin Ganye	Project information	31 Dec 2019
100MW Dangote-black Rhino solar power plant	Project information	31 Dec 2018
50MW solar plant in Jigawa State	Project information	31 Dec 2018
100MW (dc) Nova Scotia solar power plant	Project information	31 Dec 2018
Solar Nigeria Program	Project information	31 Dec 2020
30MW/100MW solar power plant	Project information	31 Dec 2020

5. Recommendations

As the policy and legal framework in Nigeria in support of solar energy is lacking in some ways, first, Nigeria must strengthen the existing policies in order to secure a noticeable share in the global solar power. There is a need for Nigeria to build and implement a full RE policy and legal framework irrespective of its vested interest in the crude oil and gas market because of the global tilt towards RE energy. There is also a need for Nigerian academia and public research institutions to build capacity in computational simulation of the thermal-mechanical processes/effects induced by Nigerian environment on solar PV modules. This will help to establish thermal, material/manufacturing, geometric and accelerated lifetime testing standards for domestic production of solar PV that are suited to the local environment. This recommendation is hinged on the fact that local consumers are only attracted to durable products which guarantee value for money. Nigeria - like other African countries, such as Ghana, Angola, Swaziland and Rwanda - must set a target for the share of its population with access to clean cooking. A hundred percent target by 2030 is recommended, and this recommendation can possibly galvanize interest in solar cooking and heating/cooling applications for which the technology is fully established in Nigeria. Nigeria's RE legal and policy documents must be revised to exclude uncertainties which discourage local and foreign investors. Nigerian policies on capital subsidy, grant, or rebate for solar power investments must be furnished with obligatory quantitative figures. There is the need for public investment to surmount perceived high capital costs of the solar power sector (especially for rural electrification) even in the face of the plans for full privatization of the power sector in Nigeria.

References

Aliyu, A. S., Dada, J. O. and Adam, I. K. (2015), 'Current status and future prospects of renewable energy in Nigeria,' *Renewable and Sustainable Energy Reviews*, 48: 336–346.

Bradford, T., (2006), *Solar revolution: the economic transformation of the global energy industry*. MIT Press Books (Vol. 32).

British Petroleum (2014), *BP Statistical Review of World Energy* (63rd ed.).

British Petroleum (2018), *BP Statistical Review of World Energy* (67th ed.).

Bugaje, I. (1999), 'Remote area power supply in Nigeria: the prospects of solar energy,' *Renewable Energy*, 18(4): 491–500.

ECOWAS Centre for Renewable Energy and Energy Efficiency, ECOWAS renewable energy Policy (2013).

Fadare, D. A. (2009), 'Modelling of solar energy potential in Nigeria using an artificial neural network model,' *Applied Energy*, 86(9): 1410–1422.

Federal Ministry of Environment (2015), *Nigeria's Intended Nationally Determined Contribution*. Abuja.

Federal Ministry of Power (2015), *National Renewable Energy and Energy Efficiency Policy (NREEEP)*. Abuja.

Giwa, A., Alabi, A., Yusuf, A., and Olukan, T. (2017), 'A comprehensive review on biomass and solar energy for sustainable energy generation in Nigeria,' *Renewable and Sustainable Energy Reviews*, 69: 620–641.

Global Tenders (2018), 'Energy, Power & Electrical Tenders Proposals Projects Bids - Nigeria'.

GOPA-International Energy Consultants GmbH (2014), *The Nigerian Energy Sector - an Overview with a Special Emphasis on Renewable Energy, Energy Efficiency and Rural Electrification*. (K. Ley, J. Gaines, & A. Ghatikar, Eds.) (1st ed.). Deutsche Gesellschaft für Internationale Zusammenarbeit (GIZ) GmbH; Nigerian Energy Support Programme (NESP); Federal Ministry of Power.

GOPA-International Energy Consultants GmbH (2015), *The Nigerian Energy Sector - an Overview with a Special Emphasis on Renewable Energy, Energy Efficiency and Rural Electrification*. (K. Ley, J. Gaines, & A. Ghatikar, Eds.) (2nd ed.). Deutsche Gesellschaft für Internationale Zusammenarbeit (GIZ) GmbH; Nigerian Energy Support Programme (NESP); Federal Ministry of Power.

National Population Commission (NPC) [Nigeria] and ICF International (2014), *Nigeria Demographic and Health Survey 2013*. Abuja, Nigeria, and Rockville, Maryland, USA.

National Renewable Energy Laboratory (2018), 'Concentrating Solar Power Projects by Country'. Available at: https://solarpaces.nrel.gov/by-country (Retrieved November 13, 2018).

OFFGRID NIGERIA (2017), 'Nigeria includes solar panels, batteries, others in revised pioneer status law'. Available at: http://www.offgridnigeria.com/nigeria-includes-solar-panels-batteries-others-revised-pioneer-status-law/ (Retrieved November 22, 2018).

Ozoegwu, C. G. (2018), 'New temperature-based models for reliable prediction of monthly mean daily global solar radiation,' *Journal of Renewable and Sustainable Energy*, 10(2), DOI: 10.1063/1.5006805

Ozoegwu, C. G. (2018), 'The solar energy assessment methods for Nigeria: The current status, the future directions and a neural time series method,' *Renewable and Sustainable Energy Reviews*, 92: 146–159.

Ozoegwu, C. G., Eze, C., Onwosi, C. O., Mgbemene, C. A., and Ozor, P. A. (2017), 'Biomass and bioenergy potential of cassava waste in Nigeria : Estimations based partly on rural-level garri processing case studies,' *Renewable and Sustainable Energy Reviews*, 72: 625–638.

Ozoegwu, C. G., Mgbemene, C. A., and Ozor, P. A. (2017), 'The status of solar energy integration and policy in Nigeria,' *Renewable and Sustainable Energy Reviews*, 70: 457–471.

REN21 (2014), *ECOWAS Renewable Energy and Energy Efficiency Status report 2014*. Paris.

REN21 (2015), *Renewables 2015: Global status report*. Paris.

REN21 (2017), *Renewables 2017: Global status report*. Paris.

REN21. (2018), *Renewables 2018: Global status report*. Paris.

Swartman, R. K. and Ogunlade, O. (1967), 'Solar radiation estimates from common parameters,' *Solar Energy*, 11(3–4): 170–172.

The International Energy Agency (2013), *Renewable Energy Outlook; World Energy Outlook.*

Tian, Y. and Zhao, C. Y. (2013), 'A review of solar collectors and thermal energy storage in solar thermal applications,' *Applied Energy*, 104: 538–553.

U.S. Energy Information Administration (EIA) (2016), *Country Analysis Brief: Nigeria.*

World Energy Council (2013), *World Energy Resources: 2013 survey*, World Energy Council, 11.

Chapter 11

Solar Energy Policy and Technology Development in Kenya

Benard O. Muok[1]

1. Introduction

In sub-Saharan Africa, electrification efforts outpaced population growth for the first time in 2014 (IEA, 2017). Despite this momentum, around 675 million people – 90% of them in sub-Saharan Africa – may remain without access to electricity in 2030 (down from 1.1 billion today), and 2.3 billion could continue to rely on biomass, coal or kerosene for cooking (from 2.8 billion today). Household air pollution from these sources is currently linked to 2.8 million premature deaths per year. Furthermore, several billion hours are spent collecting firewood for cooking, mostly by women, that could be put to more productive uses (IEA, 2017).

Kenya through its long-term development blueprint Vision 2030 has set itself an ambitious target aimed at transforming the country to a newly industrializing, middle-in-

1 Director, Centre for Research, Innovation and Technology (CRIT), Jaramogi Oginga Odinga University of Science and Technology, P.O. Box 210-40601, Bondo, Kenya. Email: bmuok@yahoo.com / bmuok@jooust.ac.ke

come country. To achieve this, the government recognizes the importance of energy as one of the key enablers to achieve the vision. The country through its energy investment plan had proposed to increase its generation capacity to 5000 MW by 2016. Yet, despite the effort, the current generation capacity remains at about 2,150 MW to serve its population of more than 43 million, causing a serious constraint on economic growth.

Renewable energy has the potential to improve the generation capacity and contribute to the energy access. For example, Kenya is believed to possess more than 7,000 MW of undeveloped geothermal energy resources in the Rift Valley. Wind, solar and biomass energy are also significant potential sources for power generation. Since the publication of sessional paper number 4 of 2004 and Energy Act 2006, which prioritized the development of renewable energy sources, Kenya has made great strides in exploiting its renewable energy potential. Currently, geothermal contributes close to 30% of the total energy generation capacity, while Lake Turkana Wind power Plant is expected to add up to 300MW when it is finally commissioned and Kinangop wind plant will contribute up to 60MW. However, other renewable energy sources such as solar and small hydros are yet to be adequately exploited to contribute significantly at the national scale.

Despite the slow exploitation at the national level, solar market in Kenya is one of the fastest growing markets in sub-Saharan Africa. The rapidly developing solar market in Kenya is buoyed by conducive market environment and the falling cost of solar technology in the world market. According to the World Energy Outlook 2017, the rapid deployment and falling costs of clean energy technologies are key factors in clean energy market. For example, in 2016, growth in solar PV capacity was larger than for any other form of generation; since 2010, costs of new solar PV have come down by 70%, wind by 25% and battery costs by 40%.

2. Stakeholder analysis

The actors in the solar market, classified through a Market Mapping Approach (MMA), are: (i) Market or value-chain actors; (ii) Players involved in regulating or enabling environment; and (iii) Players providing supporting services (Griffith & Mike Albu, 2005).

Main value-chain actors: These include primary producers, secondary producers, large-scale processors, small-scale processors, wholesalers, distributors, retailers, local market, niche market, institutional customers, and local customers, etc.

Enabling environment: In this category are those involved in regulation and policy issues – trade standards, taxes and tariffs, land regulation, financial policies, business regulation, etc.

Supporting services: These consist of main supporting services such as market information, professional and business associations, financial services, legal services, extension services, and research and development.

Figure 1 shows a detailed solar market mapping in Kenya.

Figure 1: Mapping of Detailed Solar Market in Kenya

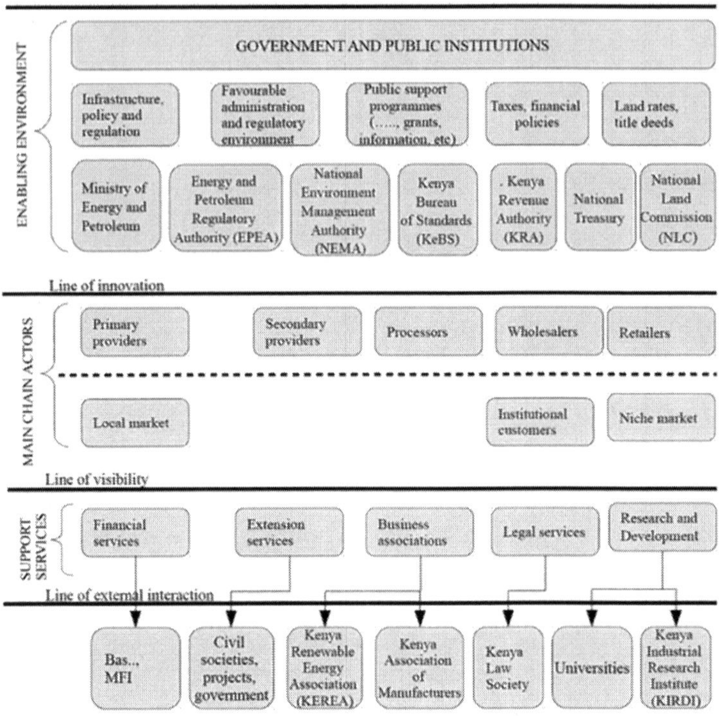

Source: Author

3. Status of solar technology development and manufacturing in Kenya

Despite the rapidly developing market in Kenya, technology development has not picked the same trend. Almost all the solar products in Kenya are imported mainly from China, India and Europe with the former currently dominating the market. At present, there are only two companies involved in some form of production. Solinc East Africa based in Naivasha is known to locally assemble solar panels while Sunlight and Mibawa designs products in Kenya but manufacturing is done in China.

The national policy environment recognizes local manufacturing as a key driver for economic development in Kenya's vision 2030 whose overall goal is to increase the contribution of manufacturing to the national GDP by at least 15% per annum and help propel the country towards becoming Africa's industrial hub. Despite Kenya's manufacturing sector being the strongest in the East African Community (EAC) region, it has not expanded at the expected rate. The key challenges facing the Kenya manufacturing sector include unreliable, costly and unsustainable energy; limited access to long-term credit for small and medium industrial enterprises; inadequate investment in industrial knowledge base, innovation, research and development; mismatch between available technical skills and market demands from the industry; and high cost of land for investment among others. In addition, local manufacturing is hampered technology ownership and tight intellectual property regime controlled by the developed countries.

Solar technology development is key in long-term solar market development in the country. Studies elsewhere have shown that locally developed technologies are more acceptable since they take into consideration of the socio-economic and cultural needs of the population (Lundvall et al., 2009). Furthermore, with weak institutions in the third world countries and especially weak capacities among the regulatory agencies, cases of counterfeit products entering the market is a constant problem. The effect of counterfeit is not only felt in the distortion of the market but also affects consumer confidence because of lack of warranties and defective products which do not last long in use.

4. Solar market

Due to poor records and data by regulatory authorities, it is difficult to accurately tell the number of products in the market. However, a recent report shows that up to 42 manufacturers have had their products tested and approved by the Lighting Africa global test standards and issued with a certificate of conformance. The following are some of the technologies in the solar market in Kenya:

Institutional solar system: These systems have a capacity ranging from 500Wp to 2kWp, but are largely determined by demand and rural electrification master plan. The institutional solar systems were the first in the technology in Kenya. Their entry dates to over 40 years ago. The government and civil societies, mainly missionaries, supported the technology entry. They were mainly installed in missionary schools, hospitals and public institutions in remote areas. To-date there are over 4000 standalone solar PV installed in schools and about 500 earmarked for installation all over Kenya. They are primarily used for lighting, refrigeration, broadcasting in remote stations, tourist facilities, cathodic protection of oil pipelines and water pumping.

Solar home systems (SHS): The last decade has seen the solar market in Kenya dominated by solar home systems. SHS are mainly driven by an individual decision to acquire and install solar PV. Such systems have a capacity of between 10 and 100Wp and have been used for charging wet batteries and lighting. Total installation capacity was estimated at 20MWp cumulative in 2013 (KCIC). The actual statistics, however, is not able to be obtained due to the increasing adoption rate and poor records from the regulatory authorities. Muok, Makokha, & Palit (2015) provide an estimate of 300,000 solar home systems. It is estimated that the overall solar market has increased by more than 100 per cent in Kenya in the last decade (Muok & Makokha, 2017). The transition towards market-based diffusion that has dominated the last few years has mainly been driven by conducive enabling policy and regulatory frameworks, comprising exemptions from value-added tax (VAT) and import taxes, Solar Regulation 2012 and innovation in the market including financing systems.

Pico-solar: International Energy Agency (IEA) defines "Pico Solar PV Systems" as small PV systems with their PV panel capacity usually a few Watt peak, but can be as small as 0.3 Wp or up to 10 Wp (IEA, 2013). These are portable systems and the battery that power 1-4 lights. Systems with extra components such as radio and mobile phone are increasingly becoming popular in rural areas where such services are required. In the past years, pico solar has become popular and the sales are rapidly increasing. Major distributors include D.light, Barefoot, Sun-King and M-Kopa amongst many others. The products have become popular in rural areas. However, there is a huge information gap regarding the level of market penetrations. Lighting Africa's sales results, collated from manufacturers of quality tested products, indicate that over 2.2 million units of solar lighting products have been sold in Kenya since 2009 and expected to have grown to over 2.7 million by 2016 (Lighting Africa 2016).

Mini-Grids: Mini-grids are among the new entries to the solar market in Kenya. Mini-grids connect villages and towns far from national grid networks. They range between 5kW and 1MW. There are about 12 mini-grids driven by donors and NGOs in Kenya (KCIC, 2017). A classic example is Kitonyoni mini-grid with a capacity of 13.5kW. The mini-grid in Makueni County was financed by the UK government and commissioned in 2012 by the Sustainable Energy Research Group at University of Southampton. The Kitonyoni Solar mini-grid is managed by Makueni County Solar Energy Co-op Society Ltd, which is owned and managed by the community. Other mini-grids are SOS children home (60kWp), Mandera (300kWp), Hola (60kWp), Helwak (50kWp), Nanyuki flower farm (60kWp), Azimuth flower solar project (72kWp), Kericho County solar plant project (1MWp), Powerhive East Africa Ltd, 4 mini grids (80kWp), etc.

Market and street lighting: The devolved government system has enabled efforts to bring lightings to rural areas especially rural markets to provide security and help extend business hours. Mainly, the county governments drive this segment. Such a system ranges from 0.2kW to 15kW in capacity.

Grid connected solar: This market segment has largely been driven by standardized power-purchasing agreements and feed-in tariffs (FIT). Kenya first developed its FiT in 2008, which was revised in 2010 and 2012. Small scale grid-connect ranges from 0.5MW – 10MW, whereas large scale from 10MW – 40MW. So far, grid-connect is dominated by small-scale actors having a capacity of 0.6MW with over 1000 players in Kenya (KCIC, 2017). Currently, the largest ones are the Strathmore University with a capacity of 0.6MWp, Garden City 0.85MWp and United National Environment Programme (UNEP).

5. Business Models

The main actors in the solar market chain include manufacturers, distributors, retailers, resellers and stockists. Four delivery models have been identified (KCIC, 2017). These include: (i) Commercially led approach where distributors and dealers/stockist develop market typically relying on cash sales; (ii) Models managed by various stakeholders and work on credit basis; (iii) Utility models mainly run by Kenya Power company distributing in remote areas, mainly in northern Kenya; and (iv) Donor grant based models, which are typically used for institutions, highly managed and structured. Of these, the first three are mainly hybrid systems.

As a result of the global trend towards solar energy, driven by the lowering prices, structure of the industry, nature and role of producers and distributors, and the necessity to address the needs of the poor (bottom of the pyramid), it is important to look closely at the business models. This brings the existing business models to the centre of policy discussion across many countries of Africa where more than half of the populations still live below the poverty line of USD 1.2 a day. While purely business approach is a prerequisite for investors, the necessity to address the needs of the poor that forms the bulk of the unconnected population in Africa is vital for governments to achieve the sustainable development goal (SDG) 7 -- universal energy access. In Kenya the following models have been employed:

Commercial model: The commercial model is operating purely on commercial basis of a willing seller-willing buyer. This has been the case of SHS whose growth was purely commercial. The government provides a level-playing field by providing a supportive policy environment including tax incentives.

Credit distribution approach: This is where a distributor enters into a partnership with institutions such as civil society organizations, micro-finance institutions, etc. which sells the products to community-based groups on a credit basis. The credits are guaranteed by membership in the groups and it's the groups' leadership that collects money and pay to the micro-finance. Since such institutions are not for profit, the profit is reinvested in market research to understand the market needs.

Pay-As-You-Go (PAYG) system: A number of innovative models are being used. Rapid developments and innovation using mobile phone platform based models (e.g. Pay-As-You-Go (PAYG)) are rapidly revolutionizing the market while providing new avenues to innovate in designing new products which means the ever-changing market demand. The mobile phone money transfer technology has made it easier for PAYG distribution system. Most manufacturers and major distributors are moving towards this direction with registered pre-determined distributor networks (e.g. big manufacturers/ distributors like D.Light solar, and M-Kopa which is run by Safaricom mobile provider).

It is clear that a lot of studies still need to be done on the delivery models. Some schools of thoughts argue that the instalment based on PAYG, whose main aim is to provide access to the poor may not be as cheap as originally thought when you consider the total amount paid by the users through instalment (Muok, 2018).

Community energy centres: Mainly supported by projects, community energy centres operate like an energy kiosk. It consists of setting a 250W power station or larger depending on the demand. Community groups often run the centre and households use the services at an agreed cost. Normally, the cost is set to be slightly lower than kerosene if the purpose is for lighting. People use it for recharging batteries of lanterns, larger batteries catering for some lights and TV/radio as well as mobile phones. Apart from those, the energy centre provides ICT services such as Internet, photocopying and printing. Depending on the capacity it may also be used for small rural enterprises such as hair clipping, low power (1hp) grain milling, small electric poultry hatcheries, milk cooling as well as the provision of village entertainment through satellite TV.

6. Policy environment for clean cookstoves development in Kenya

With a relatively advanced solar market, Kenya has made significant efforts to promote the adoption and diffusion of solar products as part of the wider objective of energy access, especially in rural areas. The government of Kenya worked closely with stakeholders including donors, private sector and researchers to provide an enabling policy and regulatory environment for the solar market. In 2014, Kenya lifted VAT charges on all solar equipment imported into the country. This nullified a 16% VAT rate brought in 2013 on imported solar products. Solar market actors and campaigners in Kenya welcomed the lifting of the tax as it reduced the cost to consumers and aids access to everyone buying solar products. However, there were also concerns over the impact of the move on Kenya's domestic solar manufacturing sector while others argued that it would open floodgates for cheap counterfeit products.

Four years down the line, the outcome of this positive step is a drastic reduction on the cost of solar product thus boosting adoption and diffusion as well as enhancing further growth of solar market in the country. The efforts by the government show the sensitivity of the government to call for policy reforms that produce an enabling environment for a well-established and sustainable clean energy. These recent actions are fundamental to pursuing a universal access to energy by 2030 as specified in the Sustainable Energy for All Action agenda, achieving the Sustainable Development Goals and implementing Kenya's Nationally Determined Contribution to addressing climate change. These positive policy changes are not only vital for the sustainable growth of clean energy market in Kenya but also a good learning experience for other countries.

Policies supporting Clean Cookstove in Kenya

Kenya has witnessed a significant transformation in the policy arena with the development of many policies and regulatory frameworks since the promulgation of the 2010 Constitution. The section below briefly outlines the specific government policies that are relevant to the development of solar market in Kenya. These policies, strategies and regulatory frameworks include the 2010 Constitution, the Vision 2030, the Second Medium Term Plan (SMTP), National Climate Change Action Plan (NCCAP), National Energy and Petroleum Policy, and the Green Economy Strategy and Implementation Plan 2016 – 2030.

The 2010 Constitution: The current Constitution which was promulgated on the 27th of August 2010 takes supremacy over all aspects of life and activity in the Country. With regard to the environment, Section 42 of the Constitution states as follows:

'Every person has the right to a clean and healthy environment which includes the right:

- To have the environment protected for the benefit of present and future generations through legislative and other measures, particularly those contemplated in Article 69;
- To have obligations relating to the environment fulfilled under Article 70;
- In Sections 69 and 70, the Constitution has *inter alia* identified National Obligations in respect of the environment and Enforcement of Environmental Rights respectively;
- For example, according to Section 69 (1): The State shall ensure sustainable exploitation, utilization, management and conservation of the environment and natural resources, and ensure the equitable sharing of the accruing benefits.

Generation and use of solar energy are among the technologies that are considered to be carbon neutral thus contributing to the environmental obligation of the constitution.

Kenya Vision 2030: The Vision 2030 is Kenya's long-term development blueprint, which is implemented through a series of five-year medium-term development plans with the third Medium Term Plan (SMTP) 2018- 2022 currently under preparation. The Vision 2030 targets a transformation of Kenya into a middle-income country with the capacity to provide a high quality of life to all its citizens by 2030. The Vision 2030 recognizes the central role of energy to the economic, social, and political development of the country. It provides a policy framework for "cost-effective, affordable, and adequate quality energy services" on a sustainable basis over the period 2004-2023.

The promotion of clean cookstove development in Kenya is seen as an important intervention to fulfilling Vision 2030's goal of enhancing access to clean energy. The Clean

Energy NAMA directly addresses multiple articles under the Constitution's Bill of Rights in the form of indirect economic-social-environmental co-benefits associated with healthcare-related particulate emissions and a clean and healthy environment (Articles 42, 69, and 70). It supports building a clean, secure and sustainable environment that cuts household air pollution by reducing the amount of wood burning. It also directly addresses Vision 2030's stated concern about the relationship between poor air quality due to wood reliance and upper respiratory infections.

Third Medium Term Plan (MTP3) 2018-2022: The Third Medium Term Plan (MTP3) 2018-2022, which succeeds the Second Medium Term Plan (SMTP) 2013-2017 underscores the need to protect the environment and building resilience to climate change (MTP3). Under the SMTP, there was a notable achievement in the area of environmental management and protection. The goals of the SMTP align with the Sustainable Development Goals (SDGs) – to achieve economic development without destroying the environment, reducing extreme poverty, achieving health and wellbeing of all Kenyan citizens, reducing human-induced climate change with sustainable energy. The MTP3 also calls for an increase in the share of energy generated from renewable energy sources. The plan seeks to offer incentives to attract both domestic and foreign investment including increased reliance on Public Private Partnership (PPP) arrangements in implementing programmes and projects. It, therefore, provides an enabling platform for the development of solar energy as a means to contribute towards a clean energy provision and environmental protection in Kenya.

National Energy and Petroleum Policy 2015: The National Energy and Petroleum Policy (NEPP) draft stipulates that Kenya's overall energy and petroleum policy is to ensure affordable, competitive, sustainable and reliable supply of energy to meet national and county development needs at low cost while protecting and conserving the environment. The NEPP draft is the result of an effort to harmonize different provisions under different legal documents governing the energy sector such as the Sessional Paper No. 4 (2004), the Energy Act 2006, the Geothermal Resources Act Nº.12 (1982), the Petroleum Act (Chapter 308 of the Laws of Kenya), and also other legal frameworks and regulations, administrative procedures, government guidelines related to energy development.

The policy emphasises the commitment of development of renewable energy. The policy also raised the serious health-related problems such as Upper Respiratory Tract Infections caused by increased indoor air pollution as a result of the use of wood fuels and kerosene in the households. It also highlights the urgency to move consumers from the consumption of kerosene and wood fuel towards more efficient renewable energy solutions (GOK, 2015 p. 95).

The Energy Bill 2015: The Energy Bill 2015 is a good response towards the recognition of the changing environment of energy regulation in Kenya. It recognizes the different sources of renewable energy and the creation of the corresponding licensing and regulatory agencies. It makes provision for the establishment of the plethora of regulatory bodies such as Energy Regulatory Authority, National Electrification and Renewable Energy Authority, Rural Electrification and Renewal Energy Corporation, Energy and Petroleum Institute, Energy and Petroleum Tribunal, etc. With the addition of so many regulatory bodies, the 2015 Bill fails in trying to streamline energy regulation to ensure that red tape is eliminated. While each entity has its own legal personality, there is a need to have a homogenous set of regulatory bodies with similar corporate structures and different functions. This Bill and others make such provision.

"(l) to undertake feasibility studies and maintain data with a view to availing the same to developers of renewable energy resources; (m) develop and promote, in collaboration with other agencies, the use of renewable energy and technologies, including but not

limited to biomass, biodiesel, bio-ethanol, charcoal, fuel-wood, solar, wind, tidal waves, mini-hydropower, biogas, cogeneration and municipal waste..." The Bill makes provision that any undertaking or works under it shall be in compliance with the local content provisions. The release of the 2015 Bill was coupled with the release of the draft "Energy (Local Content) Regulations, 2014" (the "Regulations"). Local content refers to the preference given to Kenyan citizens with suitable skills and training in matters of employment governed by the 2015 Bill. It also entails the use of certain goods and services made in Kenya and in the specific county where an energy project is being implemented. The Energy Regulatory Commission requires foreign project sponsors to have local offices where procurement, project management and implementation decision making will occur to the satisfaction of the Commission. In addition, any potential license applicant must also file a "Local Content Plan" which should generally contain plans for employment, training and succession, research and development, technology transfer, legal services, and financial and insurance services. The bill has elaborate provision for the development of renewable energy.

The Feed-in Tariff (FiT): Feed-in-Tariffs Policy (FiT introduced in 2008 and revised in 2010 and 2012) has been formulated to promote renewable energy solutions (including wind, biomass, small hydro, geothermal, biogas and solar and municipal waste energy). Under the FiT system, investment security and market stability for investors in electricity generation from renewable energy solutions are provided whilst encouraging private investors to operate their power plants prudently and efficiently to maximize returns. Though the FiT has seen a number of investments in renewable energy, experts are of the opinion that the rates are low compared to the international rates.

- *National Climate Change Action Plan, 2013*: Kenya's National Climate Change Action Plan guides the transition of the country towards a low carbon climate resilient development pathway. It designates improved cookstoves as one of the country's six top Priority Mitigation Programmes (NCCAP, 2013) and identifies the promotion of improved cookstoves as a priority intervention. This is a shift to support the Government's efforts to reduce over-reliance on fuelwood, reduce deforestation and increase access to clean reliable energy. It recognizes the considerable social, economic, and healthcare-related co-benefits associated with improved cookstoves, especially for women and children – including reducing time to collect fuelwood, reducing indoor air pollution, and potentially introducing cost savings to households.

- *Green Economy Strategy and Implementation Plan (GESIP) 2016-2030*: The GESIP is Kenya's blueprint in advancing towards a low-carbon, resource efficient, equitable and inclusive socio-economic transformation. The plan builds upon Kenya's commitment to a Low-Carbon Development Pathway and represents an advancement of this commitment to integrate resource use efficiency and minimizing environmental impacts into Kenya's economic development. Even though the plan does not specifically mention clean cooking solutions as a means to pursuing a low carbon pathway, the plan clearly reiterates the need for eco-innovation and technologies that address indoor air pollution, increase resource use efficiency, ensure equity and social inclusion. These principles directly speak to the critical functions and contribution of clean cooking solutions to Kenya's green economy agenda.

Nationally Determined Commitment (NDC)

Following the signing of the Paris Agreement, Kenya submitted its Intended Nationally Determined Commitment. In the commitment, energy is a key target. The NDC target to achieve proportionate emission reduction in the energy sector is 15.4 MtCO2e (9.32 MtCO2e from electricity generation and 6.09 MtCO2e from energy demand). This will be mainly achieved through the development of renewable energy resources and energy efficiency technologies.

Big Four Agenda: President Uhuru Kenyatta in his address to Kenyans during the 54th Jamhuri Day celebrations on December 12, 2017, unveiled what he termed as "The Big Four," the agenda that defined the Government's priorities and development path for the 2018–2022 planning cycle. The Big Four Agenda, which aims at accelerating economic growth, focuses on manufacturing, food security and nutrition, and providing universal health coverage and affordable housing. At the global level, the Big Four Agenda is effectively aligned to the 2030 Agenda for Sustainable Development, upon which the SDGs are anchored. All the Big 4 Agenda require sustainable and reliable energy. Solar power is seen as one of the viable renewable energy alternatives.

Environmental Management and Coordination (Fossil Fuel Emission Control) Regulations 2006: These regulations are described in Legal Notice No. 131 of the Kenya Gazette Supplement no. 74, October 2006. They are the internal combustion engine emission standards, emission inspections, the power of emission inspectors, fuel catalysts, licensing to treat fuel, cost of clearing pollution and partnerships to control fossil fuel emissions used by the contractors. The fossil fuels considered are petrol, diesel, fuel oils and kerosene. Solar power is seen as one of the viable renewable energy alternatives. Particularly, solar is seen as potential alternative to kerosene which is used widely for household lighting.

The Kenya Grid Code Policy: The Kenya Electricity Grid Code of March 2008 was developed to facilitate the Kenya Government's initiatives in power sector reforms as discussed above and is intended to be a technical guideline to accompany the PPA. The Code deals with matters in relation to the industry, which requires regulation, such as the supply and sale of electricity to consumers, distribution system operation and retail metering.

The objectives of the Grid Code relating to renewable energy feed-in are to provide for:

- a regime of regulation of the industry;
- a set of industry-oriented rules authorised by the Commission governing industry operations;
- power system security, network connection and access and network services pricing;
- a framework for dispute resolution;
- adequate sanctions in cases of breaches of the Grid Code;

And as pertains to technical and industry operations, to guide on:

- rules for the dispatch process and for the purposes of producing bills and credit notes;
- detailed operational requirements, including power system operations and power system security, emergency operations, metering, including metering obligations of the distribution network service provider, and maintenance scheduling;
- terms and conditions of access and technical standards that will apply for a connection to the network;
- the methods to be used for pricing network services;
- regulation of the supply of electricity to or from the distribution network of service providers' distribution systems.

Solar PV Systems and Solar Water Heating Regulations: The Energy (Solar Photovoltaic Systems) Regulations of 2012 requires that the design, installation, repair and maintenance of a solar PV system should be in accordance with the relevant national standards. This ensures that system design and specifications take into account the electric energy needs and safety of the local users. For monitoring and quality control purposes, solar

PV system technician or contractor must issue an installation completion certificate upon the commissioning of the solar PV system, showing as a minimum, the date of installation, details of the person installing, details of the owner, the location, capacity and warranty. The role of the owner is then clearly stipulated as to use and ensure the maintenance and carry out repairs required to keep the installation in good and efficient working condition after receiving necessary documents as above from the installer. The warranty for the components in the solar PV system and the PV Installation shall be clear on the set-out periods and consumer devices, which incorporate the use of solar PV systems.

The licensing regime is as follows:

- Licensing of solar PV system technicians so that a person shall not design or install any solar PV system unless he/she is licensed by the commission. Classes of these licenses vary with qualification and scope of work as follows:

    i. Class T1 licence entitles the holder to carry out solar PV system installation work for small systems or single battery DC system of up to 100 Wp.

    ii. Class T2 licence entitles the holder to carry out solar PV system installation work for medium systems or multiple batteries, which may include an inverter.

    iii. Class T3 licence entitles the holder to carry out solar PV system installation work for advanced, including grid connected and hybrid systems.

- Licensing of PV manufacturers: This ensures that a person shall not engage in the business of manufacture of any solar PV system and components unless he/she applies for and obtains a licence from the Commission.

- Licensing of solar PV importer, vendor or contractor: This stipulates that a person shall not import, distribute, promote, sell or install any solar PV system unless he/she is licensed by the Commission as a vendor. The following classes of licences shall apply:

    i. Class C1 licence entitles the holder to carry out design and installation work for solar PV systems, with a class T3 technician in employment.

    ii. Class V1 licence entitles the holder to design, distribute, promote, sell or install solar PV systems, with a class T2 technician in employment.

    iii. Class V2 licence, which shall entitle the holder to manufacture or import solar PV systems or components, with a class T2 technician in employment.

The Energy (Solar Water Heating) Regulations, 2012: Solar Water Heating regulation was made in pursuit of the Energy Act, 2006 to enforce installation of solar water heater on all premises within the jurisdiction of a local authority with hot water requirements of a capacity exceeding 100 litres per day. The enforcement clause provides a grace period of five years from the date of coming into force after which all existing premises qualifying under the above-stipulated condition must comply. But even before the expiry of the grace period, the market is seen to be picking up not necessarily for the sake of complying but rather reaping the benefits of energy saving. There is a growing awareness, with the efforts initiated by the Kenya Association of Manufacturers (KAM) on the economic sense of practising energy efficiency with incorporation of renewable energy solutions. But on the whole, this regulation provides a convenient primer on how Government intervention through awareness and regulation can invigorate a market.

Kenya Standard KS IEC/TS 62257-9-5 for pico-solar: One of the major challenges for solar market development in Kenya is the proliferation of sub-standard counterfeit products in the market. The national regulator lacks the capacity to enforce the law to regulate the market. The World Bank supported programme, Lighting Africa, has developed a voluntary certification and labelling system that has brought some sanity to the market, but it is completely voluntary.

The government's recent effort has resulted in the pico solar standards KS IEC/TS 62257-9-5. This is a step in the right direction but more still needs to be done in creating standards for the whole market and to develop capacity to enforce the law.

6. Policy recommendations

- Limited availability of credit financing is one of the key challenges facing solar development. Setting credit schemes by the government and development agents could help to address the issue. In particular, the proposed Climate Fund could have a provision for credit financing for value chain actors in the solar market.
- Further policy incentives could help to spur investment in the solar market. These could include incentives such as favourable tax holidays on investment on solar products.
- There should be a concerted effort to create awareness and capacity building of the rural population and actors in the value chain including consumers, technicians and entrepreneurs.
- There is the need to build capacity of the regulatory bodies not only to develop policy and standards but also on enforcement. This will help to reduce counterfeit products in the market hence creating confidence in the market.
- Finally, there is a need to develop a FiT for mini and micro grids below the 500kW cap as it stands today.

References

International Energy Agency (IEA) (2017), *World Energy Outlook Report 2017*.

International Energy Agency (IEA) (2013), *Pico Solar PV Systems for Remote Homes*.

Lighting Africa (2016), *Off-grid Solar Market Trends Report*. Available at: https://www.lightingglobal.org/resource/off-grid-solar-market-trends-report-2016/

Lundvall, B.-Å., Vang, J., and Joseph, K.J. (2009), 'Innovation Systems Research in Developing Countries', In: Lundvall, B.-Å., Vang, J., Joseph, K.J. and Chaminade., C. (eds.), *Handbook of Innovation Systems and Developing Countries*. Cheltenham: Edward Elgar Publishing, 1(32).

Muok, B. O. (2018), 'Developing an Innovative and Sustainable Market-Based Systems for Enhancing Diffusion of Clean Cooking Solutions in Kenya', *International Journal of Sciences*, 4(6): 57–67.

Muok, B.O., Makokha, W. B. and Debajit, P. (2015), *Solar PV for enhancing electricity access in Kenya: What policies are required?* The Energy Resource Institute, New Delhi, India.

Chapter 12

ANSOLE DAYS:
Demystifying Solar Energy Research through Artistic Creativity

Emelda Ngufor Samba[1]

In recent years, theatre has ceased to be the prerogative of theatre halls only and has moved to unconventional venues like marketplaces, football fields, fon's palaces, prison yards and even restaurants. That was the case of the performance of *'No Bills with the Sun'*, a play created and performed by Master's students of the Performing Arts and Cinematography Section of the University of Yaounde I on the occasion of ANSOLE DAYS, 2012 in Cameroon. No Bills with the Sun, a play on the challenges and social ramifications of solar energy, was performed to a mix audience of conference participants and restaurant clients who, in addition to savouring an end-of-the-day meal, also savoured all that an African total theatre could offer; dialogue, narration, songs and dance.

The purpose of this paper, therefore, is to review the interconnectedness of arts and sciences and also the nature of the performer-audience interaction during the performance and more importantly, the complex processes of collaboration that

1 Department of Arts and Archaeology, Faculty of Arts Letters and Social Sciences, University of Yaounde 1, Cameroon. Email: amboebems@yahoo.com

built up into this production. Four main levels of interaction could be identified here, namely the University of Yaounde 1, the African Network of Solar Energy, ANSOLE, and the DAAD at the level of funding and organization, Intra-troupe collaboration, the coordinators and students of the Performing Arts and Cinematography Section, and the performer-spec-actor collaboration. These levels of collaboration, in much defined ways, influenced both the process and outcome of the performance.

The interpenetration of the arts and sciences is fast becoming the norm, especially as more and more, scientists are seeking artistic expressions to give meaning to their scientific explorations and findings (see Kirstin Shepherd-Barr's *Science on Stage: from Dr Faustus to Copenhagen,* 2006) At the social sphere, there is a dire need to transform absolutely complex and somewhat complicated scientific concepts to simpler expressive modes that could be easily understood by those for whom the final products of research works are geared. It was against this background that collaboration between scientists and performing artists was established; to produce a play on solar energy that would explore its possibilities and social ramifications. Such a play was to answer, in very simple terms, a number of questions that were imagined would preoccupy people confronted by this new concept of renewable energy; what is renewable energy/solar energy? How does it work? Why should communities and individuals consider using it? How affordable and accessible is it? These and other questions were to be raised in this production.

The theatre, be it mainstream or improvised, is basically a collaborative art form. Through a three –tier collaborative effort, a one-hour play, ***No Bills with the Sun,*** was created and performed on the occasion of ANSOLE DAYS 2012. The play tells the story of how a community came into considering the use of solar energy as an alternative to hydro-electricity. Through a compelling narrative and dramatic techniques, the audiences are carried along by the performers to witness the tragic death of an electrician in the process of trying to resolve an electrical problem. At a wedding ceremony, there is power cut and the community electrician is hired to fix the problem. In his effort to do this, he is electrocuted, and dies. His brother, on trying to rescue him is also electrocuted leaving the village with two corpses in one evening.

This incident prompts the community to start wondering if there is no less risky source of energy supply especially as population increase has led to frequent power cuts, electricity bills have become too high and there has been an increase in the death rate by electrocution. Alternatives like candles, generators and kerosene lamps are proposed but the people point out the limitations of each; while candles are limited to lighting small sections of the house, people are known to have suffocated and died from the inappropriate use of kerosene lamps. Additionally, the price of petrol is permanently on the increasing.

A member of the community remembers hearing Mummy White, a middle age woman in the community talking about '*a kind of bag that generates electricity*'. This lady is known to be so passionate about her bag that she carries it wherever she goes and charges her phone and batteries with it. The people become curious and would like to see Mummy White's bag. When finally, she appears, it is amidst jubilation, singing and dancing. She recounts that the bag had been brought to her from Germany by her son who had studied on the DAAD scholarship. In all excitement and curiosity, the community members start asking her questions about the bag, how it functions, how much it costs, if it can light an entire house, etc. The only information she can offer is that, she can charge her phone with the bag and that the bag can light small bulbs. In their frustration, the performers acting as villagers turn to the experts in the restaurant to get more information about the bag. Spontaneously, one of the conference participants, an expert in the field provides detailed information about solar energy in general and solar bags in particular.

The initial purchase of solar panels could be expensive but subsequently, no bills are paid with the sun. Another conference participant volunteered the information that in some communities, people have grouped themselves and collectively bought large solar panels that could provide electricity for all everyday uses. Having received detailed and satisfactory explanations to their concerns from the experts, the performers, accompanied by the audience, dance in celebration to this newly acquired knowledge.

The play ***No Bills with the Sun*** falls within the realm of what James Thompson (2008) calls, applied theatre, the kind of theatre applied in a given situation of bewilderment. Applied theatre becomes necessary in situations where communities, faced with extremely problematic conditions have given up hope of ever finding a solution. In this play, the community is faced with a challenging situation; constant power cut, high cost of petrol (for generators), risk of electrocution in the case of false manipulation of electricity cables, risk of suffocation from lanterns, etc. Very little help can emanate from such a community that has already exploited all the resources, both human and material, at its disposal. The people's bewilderment therefore compels them to seek assistance from without, most often away from their immediate community.

What is of capital importance in this kind of theatre is that the need is identified and analyzed by the community in question. It is only after futile attempts to resolve existing problems that outside help is sought. That way the desire for change and the transformation thereof comes from the people themselves and not as a consequence of external decisions forced down on them. In the case of ***No Bills with the Sun***, the solution came from members of the audience; researchers and experts in the field of solar energy.

As aforementioned, *No Bills with the Sun* came into being as a result of collaboration at four different levels. The first of this was the DAAD- African Network of Solar Energy, ANSOLE. Here was the question of DAAD co-sponsoring the ANSOLE DAYS 2012 event that took place at the University of Yaounde I, Cameroon campus. As part of the decision of co-financing, it was agreed that, a theatre performance on Solar Energy be done within the conference. The main objective of this play was to draw people's attention to the need for alternative energy in an era where global warming is already having its impact on the environment.

The second level of collaboration was between ANSOLE and the Performing Arts and Cinematography Section of the University of Yaounde I. The theatre coordinators decided to integrate this into the class project for the Masters I students of this section who at that time were doing the course Theatre for Conscientization and Mobilization. This level of collaboration was beneficial to both partners in the sense that, while the students had their class project partly sponsored; the conference participants of ANSOLE DAYS had a chance to have a retrospective view of most of what had been discussed during the conference, this time, from an artistic point of view.

Collaboration also took place at the level of the coordinators on the one hand and the students on the other hand. As is the case with applied theatre, a play script is created by members of the community during sessions of brainstorming, analysis, identification of main issues, narration of personal experiences, story creation, and rehearsals. Both coordinators and students began by expressing their views on solar energy, some shallow, others relatively profound. It was within this framework of listening to each other, sharing information, expressing viewpoints and extrapolating from others experiences that a story was finally built up.

The fourth level of collaboration and probably the most exciting and important was that between the performers and the audience. At the beginning of the performance, the workshop coordinator announced that members of the audience would be expected to participate actively during the performance. It therefore came as no surprise that as soon as light went off (as designed in the play) during the wedding ceremony, a member of audience spontaneously offered his mobile phone with a flashlight thereby transforming himself into what Augusto Boal (2008) calls *spec-actors*, that is, a spectator who is at the same time an actor. More transformation from passive spectators to active participant-spectators occurred when solar energy experts in the restaurant spontaneously joined the performers in the performance space in response to the latter's request for more explanations on the functioning of solar energy. The most exciting was the celebration scene where a vast majority of the audience joined the performers in singing and dancing. At this point, there was no clear demarcation between the performers and the audiences since the latter found themselves completely implicated in an issue that for them was a passion.

The passion and energy injected into this last phase of the performance seemed to me not just a celebration of the end of the play but rather a dance in celebration of the success of the conference. Also, the fact that the performance took place in a restaurant during the conference dinner gave it additional impetus. The choice of venue and the unintended outcome affirmed the assertion of Philip Bell and co. who posit that, "Clearly, informal environments can and should play an important role in science education now more than ever. Learning science in informal environments has the potential to bolster science education broadly on a national scale" (2009: 26).

Carrying out a performance on solar energy in this unconventional venue, a restaurant, not only contributed in demystifying the concept but also provided an opportunity for non-professionals in the field, restaurant clients, to encounter a rather new and complex concept. The play came across strongly as a unique learning experience for these restaurant clients who, without their consent or pre-knowledge, were offered a meal of complex scientific concepts demystified by artistic creativity and served in a bowl of total theatre; dance, song, narration and dialogue. While savouring a blend of tasty traditional dishes from Cameroon, these restaurant clients, both experts and people with very little or no knowledge of solar energy also had a chance to savour a theatre performance.

The play further demonstrated the interconnectedness between entertainment and education on the one hand and between arts and science on the other. Through the question and answer session in the play, the experts got to know the aspects of solar energy research that would be of interest to the common man on the street or better still to the users. It became clearer to these experts that while a conference is capital for scientists to methodically prove their findings, there is need for a less complex communication system that would enable consumers acquire information on aspects of the research that directly concern them; the different solar energy products, where to get them, how to use them, how much they cost, long and short term advantages and disadvantages, etc.

To conclude, it is important to highlight the double-fold advantage of this production to the main stake holders; while the experience stood out as a new and exciting alternative approach for solar energy scientists to present findings of their research to non-expert audiences, the play production process enabled the Performing Arts students, who, before the production, had a somewhat limited knowledge about solar energy. The experience further reinforced the view that because man is basically an artist, people tend to understand artistic representations of complex concepts better than they would do under more constrained circumstances.

References

Boal, A. (1979, 2000), *Theatre of the Oppressed,* The European Union: Pluto Press.

Boal, A. (2007), 'Poetics of the Oppressed', in Kuppers, P. and Robertson, G. (eds.), *The Community Performance Reader,* London and New York: Routledge.

Shepherd-Barr, K. (2006), *Science on Stage: from Dr Faustus to Copenhagen,* Princeton: Princeton University Press.

Chapter 13

African Network for Solar Energy:

Bridging Africans in Solar Energy Research

Daniel Ayuk Mbi Egbe[1]

As destiny wanted it, I was invited to replace Professor Niyazi Serdar Sariciftci, (director of the Linz Institute for Organic Solar Cells (LIOS) of the Johannes Kepler University Linz, Austria) as a keynote lecturer on "Polymer Bulk Heterojunction Solar Cells" at the International Conference on Conducting Materials (ICOCOM 2010), which was held from the 3rd to the 7th of November 2010 in Sousse, Tunisia. ICOCOM 2010 was organized among others by Professor Samir Romdhane of the Tunisian Chemical Society. One of the conference participants, Professor Teketel Yohannes from the University of Addis Ababa, Ethiopia, an African

1 Professor, Institute of Polymeric Materials and Testing (IPMT), Johannes Kepler University Linz, Altenbergstr. 69, 4040 Linz, Austria. Member of Organic Electronics Association (OE-A), and a board member of the World University Service (WUS) in Germany. International Coordinator, African Network for Solar Energy (ANSOLE) (www.ansole.org). Coordinator of BALEWARE (Bridging Africa, Latin America and Europe on Water and Renewable Energies Applications) (www.baleware.org). Email: daniel_ayuk_mbi.egbe@jku.at

expert in the area of organic solar cells, was beforehand less known by the organizers. This may be related to the linguistic and other cultural barriers existing among African scientists imposed on them from past and present external influences. Although the situation is gradually changing, it is known that most African scientists interact more with their European and American counterparts than with their fellow Africans. Beside the cultural barriers, other factors like lack of transportation infrastructures (good roads connecting different African countries or direct flights between major African cities) contribute in separating African scientists. For instance, due to the lack of direct flight between Addis Ababa and Tunis, Professor Teketel Yohannes had to fly through Dubai, which took him more than 8 hours.

Conscious of this situation while in Sousse, the desire of bridging African scientists in the field of solar energy research was born in my mind. After consulting with Prof. Romdhane and Prof. Yohannes, the three of us decided to start up a network of exchange among African scientists. Together with 6 other scientists from North Africa, we met between 10 pm and midnight of the 4th of November 2010 in a conference room of Hotel Al-Mouradi in Sousse, Tunisia to initiate an African network in the area of solar energy.

Africa has been blessed with much sun and it can solve both its energetical and environmental problems by developing the solar energy sector. This requires qualified manpower to be trained in the framework of such a network, the 9 participants of the kickoff meeting in Sousse unanimously agreed on this. The network was named "African Network on Conducting Materials for Solar Energy (ANCMSE)" in association to ICOCOM. For language reasons, I was asked by the participants to coordinate ANCMSE at the early stage of its existence.

The name "ANCMSE" was, however, found to be very exclusive(restrictive). It was dropped in favour of "African Network for Solar Energy" having the acronym "ANSOLE" during the second constituting meeting, which was held in the form of a mini-symposium exactly 3 months later, on the 4th of February 2011 at the LIOS, in Linz Austria. This second gathering was possible thanks to Professor Serdar Sariciftci, who succeeded to gather some funding to cover the expenses of most of the African attendees. Lesley Tobin of ICPC-Nanonet (http://www.icpc-nanonet.org) was very helpful to the success of the mini-symposium. With the information she provided, we were able to invite interesting participants coming from different African regions, such that the decisions taken during the symposium were binding. I am, moreover, very grateful to her for the practical help she offered, such as writing down the minutes and designing the very beautiful logo of ANSOLE. The attendance of ICTP (The Abdus Salam International Centre for Theoretical Physics) a United Nations institution based in Trieste Italy, which has long time experience working with African scientists, which was represented by Dr. Ralph Gebauer and Dr Joseph Niemela, gave some weight to ANSOLE. ICTP promised to support ANSOLE in various ways. Following countries were directly or indirectly represented at the symposium: Algeria, Austria, Cameroon, Côte d'Ivoire, Denmark, Ethiopia, Germany, Great Britain, Italy, Morocco, Nigeria, Scotland, Tunisia, South Africa, Turkey and USA.

Following scientists attended the official launching of ANSOLE:

Professor Serdar Sariciftci, Johannes Kepler University Linz, Austria
Professor Teketel Yohannes, University of Addis Ababa, Ethiopia.
Professor Samir Romdhane, University of Bizerte, Tunisia.
Dr. Getachew Wokneh Adam, University of Addis Ababa, Ethiopia
Professor Amel Romdhane, University of Tunis, Tunisia.
Professor Adelfattah Barhdadi, University of Rabat, Morocco.
Professor Izzedine Zorkani, University of Fès, Morocco.

Professor César Kapseu, University of Ngaoundere, Cameroon.
Professor Adenike Boyo, Lagos State University, Nigeria.
Professor Thouakesseh Jérémie ZOUEU, Polytechnic Institute of Yamassoukro, Ivory Coast.
Professor Malik Maaza, iThemba LABS, West Cape Province, South Africa.
Professor Mammo Muchie, SARCHI Chair, South Africa.
Professor Daniel Ayuk Mbi Egbe, University of Linz, Austria.
Dr. Ralph Gebauer, ICTP Trieste, Italy
Professor Joseph Niemela, ICTP Trieste Italy
Participants at the launching meeting selected me to coordinate the network

Almost 10 years later ANSOLE has grown to be the mouthpiece of Africa in renewable energy related issues and is strongly involved in the implementation of 17 sustainable development goals (SDGs). It is now represented in 45 African countries and 31 non-African countries. It has more than 1200 members, 90 % of them belonging to morethan 300 higher education institutions. ANSOLE has been recognized as a legal entity according to the German associations laws in the name of ANSOLE e.V.. It presently has its offices at the Wagnergasse 25, 07743 Jena, Germany.
ANSOLE has 3 main goals:

1) Foster vocational training and eduction at various skills levels
2) Foster research activities among African scientists and non-African scientists directly involved in the training and education of African students and experts
3) Promote the use of renewable energy in Africa by making it accessible to all.

ANSOLE created 3 scholarship programmes financially supported by ICTP through which it successfully graduated 1 MSc and 16 PhD candidates (2011-2016). It has organised conferences, workshops, sumposia and summer schools in Africa and Europe. To date 30 events have been organised in 16 countries. It has facilitated African students to find laboratories in Europe and has done the same to European and American students in search for hosting institutions in Africa. These are just few of its achievements in the past years....

I invite you to join this dynamic network and contribute in the development of the African continent.

www.ansole .org

Participants at the launching of ANSOLE on February 4th 2011 in Linz Austria

Participants of the mini-symposium of ANSOLE on the 4th of February 2011 in Linz, Austria.

Chapter 14

Post Scriptum

Mammo Muchie, Angathevar Baskaran & Daniel Ayuk Mbi Egbe

African countries are experiencing rapid growth of economy, population and urbanisation. By 2025 Africa's population is expected to surpass that of both India and China. Its urban population is set to grow by more than half a billion over that period, much higher than what was experienced in China since it opened its economy in the 1980s. These changes are likely to expand economic growth and infrastructure development and increase energy demand substantially. Rapid economic growth, urbanization, and changing lifestyles among both urban and rural areas in Africa are expected to double or even triple the energy demand by 2030. In addition, currently, huge population do not have access to energy services. For example, in Malawi only 7.6% of the population has access to electricity (i.e. 20% in urban and 1% in rural areas) and the country desperately needs to build its energy capacity to reach this population. In Nigeria meeting full electrification rate through conventional grid expansion covering terrains like coastal areas and remote areas is not possible. However, how Africa meets its energy needs of fast growing and increasingly urban population and population currently excluded from energy access is crucial. The major challenge is meeting the increasing energy demand in an economically and environmentally sustainable manner.

Renewable energy, especially solar, is therefore the only option for possible full electrification. Solar energy presents a viable alternative because Africa has good solar radiation resource throughout the year.

Although Africa has caused only 2 per cent of global energy-related carbon dioxide emissions, it is one of the most affected regions by the climate change. Its ecosystems are already affected disproportionately and future impacts from global climate change are expected to be even worse with serious implications for food security, migration and socio-economic development. All African countries have signed the Paris Agreement and committed to join the global effort to mitigate greenhouse gas (GHG) emissions. Furthermore, many countries in Africa are striving to transition from a resource-based economy to an industrial-manufacturing economy. Therefore, sustainable energy supply is critical to achieve this goal. Africa has a great opportunity to harness renewable energy such as wind, hydro and solar for undertaking such structural transformation of the economy. The abundance of renewable energy sources in Africa, particularly the solar energy provides a great opportunity to the governments, industry, universities and communities to chart a sustainable socio-economic development plan. Africa has deserts such as the Sahara that offers opportunities to develop large scale solar energy installations. Apart from such large-scale installations, small scale solar projects can be developed for small communities, organizations and businesses. Thus, solar energy offers a huge opportunity for Africa, both at utility-scale and off-grid levels.

Governments across Africa are devising ways to expand the share of renewable energy in the total energy mix. Several countries have already formulated policy frameworks to support the solar energy market. Also, many global solar energy firms are showing keen interest on entering the African solar market. Although the total solar deployment is low in Africa at present, it is growing fast and several large-scale projects are in the pipeline. For example, the 1.6 GW solar photovoltaics (PV) park at Benban in Egypt and the 510 MW Noor solar concentrating solar power installation in Morocco are among the largest solar projects in the world.

Although Africa is one of the most solar-rich continents in the world, it has not yet developed capabilities required to efficiently harness the solar energy mainly due to a variety of reasons such as lack of long-term strategy, weak national institutions and regulatory frameworks, lack of skills, and financial investment. Particularly, the lack of investment and technical know-how have affected the growth of the African solar energy industry. Consequently, the solar energy market in Africa is far behind other regions. For example, from the experience of South Africa, it is clear that the development of solar energy technologies is mainly hampered by weak institutional arrangement, strict regulatory framework, lack of skills, policy uncertainty and gaps, and economic climate. Similarly, although Kenya has one of the fastest and best-developed solar markets in sub-Saharan Africa, it still faces major challenges to realize its full potentials. These include slow or lack of local manufacturing capacity, inadequate financing mechanisms, still emerging business models, poor awareness especially in the rural areas, low technical capacity, and inadequate policy framework including standards and quality control.

New applications such as solar thermal cooking and power generation have good potential, but they require substantial investment. Current solar PV applications such as lighting, refrigeration, telecommunications and powering of low power electronic appliances pose several barriers including high cost of capital equipment and spares. Besides, the political and economic instability in the previous decades have left behind a poor energy infrastructure in many countries. In most African countries, the existing energy infrastructure needs to be upgraded while new infrastructure such as solar parks are being created. Many African countries such as South Africa, Kenya, Malawi, and Nigeria are taking measures to overcome the weaknesses highlighted above and build solar technological capabilities. For example, as discussed in the case of Malawi, the government is taking efforts to build solar research and technological capabilities involving government institutions, universities, non-governmental organizations and to a lesser extent the private sector.

Increasing solar energy capacity across Africa offers the best option to develop sustainable and efficient energy sector in the long-term. African countries should encourage both domestic and foreign investments in solar energy and technology to create a sustainable energy market in the long-term. Regional cooperation and coordination are crucial to achieve the efficiencies and economies of scale required to accelerate the growth of the industry. Such an approach is imperative particularly for large-scale power generation projects involving cross-border renewable resources. Adopting regional integration approach will help address issues related to trans-boundary regulatory frameworks and policies, regional infrastructure development and economies of scale and cost efficiency. This is evident from the current status of the research in solar energy in the central Africa region. Although some research laboratories and private organizations are conducting solar energy research, so far the results have failed to meet expected level. To build their research and development capabilities and positive outcomes in the African countries, it would be necessary to encourage, and finance solar energy projects based on regional integration and coordination.

Building robust solar industry and technological capabilities in Africa requires right mix of policies, incentives and multi-stakeholder collaboration at regional, national and local levels. Such integrated approach is likely to help overcome numerous challenges and bring in immense rewards such as eliminating energy poverty by enabling electricity access to rural communities, spurring industrial growth, creating solar industry ecosystem, and developing sustainable energy supply in the long-term.

Notes on Contributors

Mammo Muchie
DST/NRF SARChI Professor (Innovation & Development), Tshwane University of Technology, South Africa; Technology and Management Centre for Development (TMCD), Oxford University, UK; Associate Faculty Professor, Sussex University, UK; Adama Science and Technology University (ASTU), Ethiopia. Email: MuchieM@tut.ac.za

Angathevar Baskaran
Associate Professor, Department of Development Studies, Faculty of Economics and Administration, University of Malaya, Malaysia; Senior Research Associate of SARChI (Innovation & Development), Tshwane University of Technology, Pretoria, South Africa. Email: baskaran@um.edu.my & anga_bas@yahoo.co.uk

Daniel Ayuk Mbi Egbe
Professor, Institute of Polymeric Materials and Testing (IPMT), Johannes Kepler University Linz, Altenbergstr. 69, 4040 Linz, Austria. Member of Organic Electronics Association (OE-A), and a board member of the World University Service (WUS) in Germany. International Coordinator, African Network for Solar Energy (ANSOLE) (www.ansole.org). Coordinator of BALEWARE (Bridging Africa, Latin America and Europe on Water and Renewable Energies Applications) (www.baleware.org). Email: Daniel.egbe@ansole.org; daniel_ayuk_mbi.egbe@jku.at/

VGR Chandran
Associate Professor, Department of Development Studies, Faculty of Economics & Administration, University of Malaya, Kuala Lumpur, Malaysia. Email: vgrchandran@um.edu.my

Noël Djongyang
Department of Physics, Faculty of Science, University de Ngaoundere, Cameroon. Email:noeldjongyang@gmail.com

René Tchinda
University Institute of Technologies Fotso Victor, Bandjoun, University of Dschang, Cameroon, Email: ttchinda@yahoo.fr

César Kapseu
Department of Process Engineering, National School of Agro-Industrial Sciences, University of Ngaoundere, Cameroon, Email: kapseu@yahoo.fr

Patrice D. Dongo
Research Group on Experimental and Applied Physics for Sustainable Development, EAPhySuD, University of Dschang, Cameroon. Laboratoire des Energies Renouvelables, 'LER', Université de Dschang, B.P. 313 Dschang, Cameroon. Email: dongopat@yahoo.fr

Robert Tchitnga
Research Group on Experimental and Applied Physics for Sustainable Development, EAPhySuD, University of Dschang, Cameroon. Laboratoire des Energies Renouvelables, LER, Université de Dschang, B.P. 313 Dschang, Cameroon. Email: tchitnga@yahoo.fr

Christian T. Tchapga
Research Group on Experimental and Applied Physics for Sustainable Development, EAPhySuD, University of Dschang, Cameroon; College of Technology, Department of Electrical

and Electronic Engineering, University of Buea, Cameroon. Email: ctchito@yahoo.fr

André R. Tchamda
Research Group on Experimental and Applied Physics for Sustainable Development, EAPhySuD, University of Dschang, Cameroon, Email: rodriguetchamda@yahoo.fr

Julius Kewir Tangka
Laboratoire des Energies Renouvelables, LER, Université de Dschang, B.P. 313 Dschang, Cameroun, Email:tangkajkfr@yahoo.fr

Anaclet Fomethe
Laboratoire de Mécanique et de Modélisation des Systèmes, L2MS, Université de Dschang, B.P. 67 Dschang, Cameroun, Email: safomethe@yahoo.fr

Yacoub I. Halawlaw
Department of Physics, Faculty of Pure and Applied Science, University of N'Djamena, N'Djamena, Chad. Email: yacoubih@gmail.com

Mahamoud Y. Khaya
Institut Universitaire des Sciences et Techniques d'ABECHE, Chad. Email: mykhayal@yahoo.fr

Raoult T. Dabou
Department of Computer Science, Electronics and Automatics, National Advanced School of Agro-Industrial Sciences, University of Ngaoundere, PO Box 455 Ngaoundere, Cameroon. Email: dabour_fils3@yahoo.fr

Maxwell T. Nsiéwé
Department of Energy and Process Engineering, National Advanced School of Agro-Industrial Sciences, University of

Ngaoundere, PO Box 455 Ngaoundere, Cameroon. Email: maxwelltientcheunsiewe@yahoo.fr

Ibrahim Njukouyou
Department of in Computer Science, Electronics and Automatics, National Advanced School of Agro-Industrial Sciences, University of Ngaoundere, PO Box 455 Ngaoundere, Cameroon. Email: nikhalil2000@yahoo.fr

Fabrice Debel Eyono
Department of Agro-industrial Engineering, National Advanced School of Agro-Industrial Sciences, University of Ngaoundere, PO Box 455 Ngaoundere, Cameroon. Email: efabricedebel@yahoo.fr

Martin Kom
Groupe de Recherche sur les Technologies Médicales Adaptées aux Tropiques, Laboratoire d'Electronique et de Traitement du Signal, Département des Génies Electrique et Télécommunications, ENSP, Université de Yaoundé I, Cameroun. Email: kommart@yahoo.com

Chifundo M. Tenthani
Physics and Biochemical Sciences Department, The Malawi Polytechnic, Blantyre, Malawi. Email: ctenthani@poly.ac.mw

Esther Phiri
Electrical Engineering Department, The Malawi Polytechnic, Blantyre, Malawi. Email: ephiri@poly.ac.mw

Thomas Senganimalunje
Physics Department, The University of Malawi, Zomba, Malawi. Email: tmalunje@yahoo.com

Maturin Petsoko
FSJP, University of Yaounde 2, Cameroon. Email: maturinpetsoko@yahoo.fr

Silas K. Mulaudzi
Energy and Electricity Division, City of Tshwane, P.O Box 440, Pretoria, 0001, South Africa. Email: Silasmu@tshwane.gov.za

Rudzani Makhado
Research and Evaluation Section, Limpopo Legislature, P/ Bag x9309, Polokwane, 0700, South Africa. Email: makhador@ limpopoleg.gov.za

Chigbogu G. Ozoegwu
Department of Mechanical Engineering, University of Nigeria, Nsukka, Nigeria.
Email: chigbogug@yahoo.com & chigbogu.ozoegwu@unn. edu.ng

Benard O. Muok
Director, Centre for Research, Innovation and Technology (CRIT), Jaramogi Oginga Odinga University of Science and Technology, P.O. Box 210-40601, Bondo, Kenya.
Email: bmuok@yahoo.com / bmuok@jooust.ac.ke

Emelda Ngufor Samba
Department of Arts and Archaeology, Faculty of Arts Letters and Social Sciences, University of Yaounde 1, Cameroon. Email: amboebems@yahoo.com

Index

B

C

D

E

F

J

K

L

M

O

P

photovoltaic vi, x, 4–5, 12, 14–18, 23, 26, 36, 38, 47–48, 50–54, 58, 60, 80, 107, 111, 118–119, 125, 127–128, 131, 137, 141, 148, 154–155, 162–164, 167, 169–170, 188–190, 232

Q

R

S

T